Advances in Industrial Control

Advances in Industrial Control is a series of monographs and contributed titles focusing on the applications of advanced and novel control methods within applied settings. This series has worldwide distribution to engineers, researchers and libraries.

The series promotes the exchange of information between academia and industry, to which end the books all demonstrate some theoretical aspect of an advanced or new control method and show how it can be applied either in a pilot plant or in some real industrial situation. The books are distinguished by the combination of the type of theory used and the type of application exemplified. Note that "industrial" here has a very broad interpretation; it applies not merely to the processes employed in industrial plants but to systems such as avionics and automotive brakes and drive-train. This series complements the theoretical and more mathematical approach of Communications and Control Engineering.

Indexed by SCOPUS and Engineering Index.

Proposals for this series, composed of a proposal form (please ask the in-house editor below), a draft Contents, at least two sample chapters and an author cv (with a synopsis of the whole project, if possible) can be submitted to either of the:

Series Editor

Professor **Michael J. Grimble**: Department of Electronic and Electrical Engineering, Royal College Building, 204 George Street, Glasgow G1 1XW, United Kingdom; e-mail: m.j.grimble@strath.ac.uk

or the

In-house Editor

Mr. **Oliver Jackson**: Springer London, 4 Crinan Street, London, N1 9XW, United Kingdom; e-mail: oliver.jackson@springer.com

Proposals are peer-reviewed.

Publishing Ethics

Researchers should conduct their research from research proposal to publication in line with best practices and codes of conduct of relevant professional bodies and/or national and international regulatory bodies. For more details on individual ethics matters please see:
https://www.springer.com/gp/authors-editors/journal-author/journal-author-helpdesk/publishing-ethics/14214

John Pittner · Marwan A. Simaan

Tandem Hot Metal Rolling Mill Control

Using Practical Advanced Methods

Springer

John Pittner
Department of Electrical and Computer Engineering
University of Pittsburgh
Pittsburgh, PA, USA

Marwan A. Simaan
Department of Electrical and Computer Engineering
University of Central Florida
Orlando, FL, USA

ISSN 1430-9491 ISSN 2193-1577 (electronic)
Advances in Industrial Control
ISBN 978-3-031-58081-9 ISBN 978-3-031-58082-6 (eBook)
https://doi.org/10.1007/978-3-031-58082-6

This Springer imprint is published by the registered company Springer Nature Switzerland AG
The registered company address is: Gewerbestrasse 11, 6330 Cham, Switzerland

Paper in this product is recyclable.

Series Editor's Foreword

Control engineering is viewed differently by researchers who produce general theories and engineers who must design, calibrate, implement, and maintain control systems. Researchers very often develop algorithms for control problems with a well-defined mathematical basis, while engineers may have more immediate concerns over the limitations of equipment, quality of control, safety, security, and system downtime. The monograph series on *Advances in Industrial Control* (AIC) attempts to bridge this divide by encouraging the consideration of advanced control techniques when they offer genuine real-world benefits.

The rapid development of new control theories, techniques, and technology has an impact on all areas of engineering. This series focuses on applications that may stimulate the development of new more general control paradigms. This is desirable if the different aspects of the "control design" problem are to be explored with the same dedication that "analysis and synthesis" problems have received. The series enables researchers to introduce new ideas motivated by the challenging problems in the applications of interest. It raises awareness of the various benefits that advanced control can provide while also explaining the challenges that can arise.

This AIC series monograph is an excellent example of an application that can learn from advances in modeling and control methods but can also promote a valuable theoretical design approach. The hot-strip tandem-mill rolling problem is introduced in Chap. 1. It includes many of the control design problems arising in industrial control systems like varying transport delays, multivariable systems, nonlinearities, disturbances, and modeling uncertainties. The description of the application is very clear, which is important for understanding the complexities of the control problem. For example, it is not just mechanical parameters like thickness to be controlled but also the metallurgical properties of the rolled strip. The process model that includes details of the roll gap modeling and the loopers is described in Chap. 2.

Chapter 3 turns to the control problem and the control mechanisms/actuators. The control requirements, such as the need to maintain a nearly constant tension in the strip for good product quality, are also described. The simple and historically important "gaugemeter principle" is described as a classical approach to thickness control. The problems introduced by eccentricities in the mill rolls are described. The very useful so-called state-dependent Riccati equation (SDRE) technique, which the authors have used for different multi-input–multi-output control applications, is also introduced and various promising advanced control methods are covered briefly.

The conventional strip-mill looper and thickness control methods are described in Chap. 4. The role of feedforward control methods is considered, for example in the difficult threading phase. The multi-stand control problem is complex with many inputs and outputs and is nonlinear; however, most classical control systems were designed using linear models. The profile and flatness control problems reveal that it is not only the average thickness and tension to be controlled but also the so-called flatness (stress distribution) and thickness profile across the strip. The control of strip temperature introduces yet another dimension to the design problem. Some of the various control techniques that can be used are overviewed briefly.

Chapter 5 considers the use of advanced control systems and particularly the SDRE approach. This is a neat approach that is intuitively simple although it is based on some assumptions and approximations. The application to the tandem hot mill control problem is interesting, particularly for its robustness to disturbances and modeling uncertainties. It is an approach that is sufficiently simple that it should be acceptable to modern control plant design engineers while still offering a real potential advance in the quality of control. Fault tolerance is considered, and this is useful in hot mills where the consequences of a so-called mill wreck are significant. The final chapter—Chap. 6—concerns the various kinds of main mill drives and motors, and the converter systems, which are often an area neglected in control-theory-oriented texts.

This is a valuable text for anyone working on steel-mill controls. It provides a very clear picture of tandem hot mill control problems and suggests very realistic algorithms that use more realistic system models and offer real potential improvements. The monograph should be useful to mill-control-design engineers, researchers in dynamics and control, and to engineering students interested in difficult multivariable interacting control problems.

Glasgow, UK Michael J. Grimble
November 2023

Preface

Our work on this monograph has multiple objectives. The main objective is the presentation of some of the fundamental characteristics of the tandem hot sheet metal rolling process and its control that would be useful to interested readers of various backgrounds who might desire to improve their knowledge in this area. What is developed is a useful handy reference that describes the important concepts related to the control of this process. This is presented in a base that describes certain fundamental characteristics of the process and what significant challenges they present to the control engineer. Additionally, the content will be of interest to others of various backgrounds such as those whose activities are more directly involved with the process or have a specific interest in it. Among these would be associated academic instructors and graduate students, control practitioners, commissioning engineers, operational personnel, maintenance personnel, and others.

Another main objective is that the presentation of our recent work in this area would result in advancements in several areas of control that would result in improvements in the quality of the final output, along with a reduction in overall costs in processing, commissioning, and maintenance. To help appreciate the recent advancements resulting from our work, a brief history[1] of recent developments in control concepts and implementations over previous years is presented, plus how useful improvements in certain areas have been realized by the development of newer advanced concepts. What is presented herein is based on our earlier work for similar application to the control of the tandem cold mill, plus several additional approaches that are peculiar to the hot mill. As is shown in the chapters which follow, the methodology also includes advances which have resulted in improvements over the general conventional techniques for hot mill control and with desirable results. Additionally, these recent associated advancements in control would provide a firm basis for even further future improvements.

Some of the major issues that must be successfully handled by the process control include the following:

---

[1] A brief history of the tandem hot metal rolling process is provided in Chap. 1.

- a wide variety of products with characteristics that can change considerably between products;
- stringent customer requirements on the final product;
- a harsh mill environment that precludes many types of control and instrumentation;
- highly nonlinear strip and mill characteristics that can change considerably during normal and abnormal operations;
- long time delays that vary as the strip moves between different mill stands;
- tolerance of certain equipment failures;
- the need to provide a user-friendly environment;

plus several others as noted in more detail in Chap. 3.

In Chap. 4 several approaches in conventional control are presented, with their individual strengths and weaknesses denoted. Included are the results of the conventional methods that have been used with reasonable success in certain areas. The basic concepts that are considered and are common to many of these conventional methods are:

- use of the "gaugemeter" method to estimate the thickness of the strip as it exits a mill stand;
- application of methods to reduce the effects of mill roll eccentricities; and
- linearization of the mathematical models of Chap. 2.

Certain methods also include feedforward techniques and the particular usage of certain process measurements. Some examples are methods based on the use of automatic gauge control (AGC), multi-input–multi-output (MIMO) control, Kalman filtering techniques, and others, all of which are more fully described in Chap. 4.

In Chap. 5 four methods of well-known advanced control are presented and evaluated. These are the (algebraic) state-dependent Riccati equation (SDRE) technique, model predictive control, sliding-mode control, and H-infinity loop shaping. Of these four, the SDRE method was considered as the most desirable. This method is quite useful for the control of the tandem hot mill as it easily handles larger nonlinear MIMO applications, is user friendly, and recently has been quite popular for a large variety of applications such as the control of the tandem cold mill.

In this technique the solution of an algebraic Riccati equation (ARE) is obtained to describe the states that are feedback to the control of a nonlinear system. The system is described by a state equation whose coefficients are nonlinear functions of the states. The elements of the ARE also can be state-dependent if needed as determined by the model representing the system. Thus the nonlinearities of the system are easily handled without the need for linearization, which otherwise might be needed to be updated often during the actual operation of the process. Further, an improved method of control was determined such that an online solution of the ARE was not required at each scan of the controller during actual operation. This simplified the controller such that a fast update of the control could be easily realized as needed to handle the rapid changes in the process. Additional techniques also were developed to improve the handling of undesirable disturbances and uncertainties, which could be easily implemented to improve the overall performance. Improvements in the control of threading were not overlooked, as a method was derived to reduce undesirable overshoots in the initial raising of the looper. Other additional improvements are described further in Chap. 5.

A very basic introduction to the control of motors and drives is presented in Chap. 6. This presents an overall rounded-out description of the control of the process and helps to understand the interfaces between the process control and the drive systems.

The improvements in the control of the tandem hot metal strip control system as presented herein are considered to be of significant use in the performance of an important major industrial process. Many additional useful improvements are described in more detail in the actual text of the book. This can be of significant use to control engineers and others associated with this process.

Pittsburgh, PA, USA John Pittner
Orlando, FL, USA Marwan A. Simaan
January 2024

Contents

Notation

The notation for individual variables is as follows unless otherwise specifically noted in the associated text.

R	The field of real numbers
ϵ	Belongs to
Ξ	Defined as
A'	Transpose of matrix A
A^{-1}	Inverse of matrix A
I	Identity matrix
$\dot{x}$	Time derivative of variable x
$\|x\|$	Euclidean norm of vector x
$\|A\|_\infty$	Infinity norm of matrix A
ϵC^k	Elements of matrix or vector have continuous partial derivatives through order k
∇	Gradient; the computation of gradients is noted in Chap. 5 (Appendix)

Chapter 1
Introduction

1.1 Objectives

The rolling of hot metal strip[1] in a tandem finishing mill is a highly complex nonlinear multivariable process, with several significant time delays that change considerably during certain processing operations. This results in a major challenge to the control engineer. In response to this challenge several methods for improving the control of this process have been developed and implemented over the past fifteen years or so with reasonable success. However, there yet remains the potential in several areas for further significant discoveries that would lead to major improvements in the control strategy and ultimately improve the yield and quality of the mill output, as well as reduce the costs of design and commissioning activities.

Thus an objective of this book is the presentation of the results of our recent work that led to the discovery of several advancements in the control of this process. Included as a part of this presentation are the comparisons of our new advanced technique to other methods, which show the advantages of our new approach. Among the several areas involved are the normal operations of the mill during fully threaded and threading regimes, the maintaining of mill availability by retaining near normal operation in the presence of significant disturbances and uncertainties, and in the event of certain abnormal scenarios. The control of the loopers is given particular attention as their operation is fundamental to the reduction of excursions in strip tension and thickness which are important to improvement in the final product and the retention of the stability of rolling. Further, the structure of the controller should be friendly to the user without requiring a background in advanced mathematics. This facilitates design and commissioning efforts with reduced costs and earlier entry into production.

[1] Herein, and also in general usage, the word "strip" is understood to refer to both strip and sheet.

J. Pittner and M. A. Simaan, *Tandem Hot Metal Rolling Mill Control*, Advances in Industrial Control, https://doi.org/10.1007/978-3-031-58082-6_1

A second objective is the presentation of some of the fundamental characteristics of the tandem hot metal rolling process and its control that would be useful to interested readers of various backgrounds who might desire to improve their knowledge in this area. Thus what is presented is a useful handy reference that not only describes the concepts of our new control method but also the tandem hot metal rolling process. Thus a base is provided that helps the reader to better understand possible control issues in related applications in other newer areas of metals processing, several of which have been recently developed in the past few years.

1.2 History of Hot Metal Rolling

In the fourteenth century some of the earliest known rolling of metals occurred wherein very small rolls were used to shape cold metals such as gold, silver, or lead that were used in jewelry or works of art. The earliest known rolling of hot ferrous metals into thin sheets occurred in the seventeenth century using 2-high single stand mills (Fig. 1.1), which was mostly throughout Europe and especially in Germany, and later in the same century in England and Wales. The use of hot rolling mills increased in the seventeenth century, and their design operation was considerably improved in the eighteenth century. Some advancements are described as follows.

It was in 1728 that a patent was granted to John Payne of England for a mill of two large grooved rolls that processed bars into various shapes and forms. Such a mill was subsequently built and in operation in 1747 in Essonne, France. Later in 1759 another patent was issued to Thomas Blockley of England for a mill with grooved rolls for the polishing and rolling of metals. A bit later in the eighteenth-century mills were built with two work rolls that were independently driven at somewhat different speeds with resulting roll wear and an upturning or down turning of the workpiece. This situation was later alleviated by the coupling of the work rolls via a gearing arrangement which was described in a patent application by John Purnell of England that synchronized the rolls of a tandem mill in which the work piece was rolled in successive stands. Thus significant progress was realized in this area, and a bit later related patents were issued in England in 1776 for the tandem rolling of wire rods and in 1798 for the tandem rolling of plates and sheets. In 1783 a patent application was filed by William Playfield for a cast metal housing with screws for

Fig. 1.1 Rolling mill stands

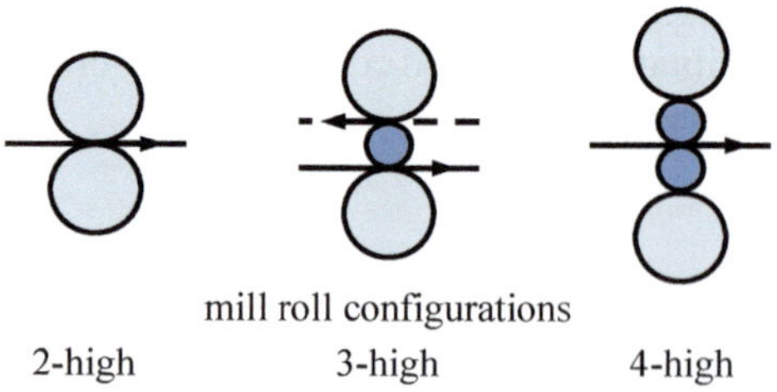

the positioning of the rolls, similar to the arrangement of mill stands that presently exist, and in 1790 a 4-high arrangement of the mill was constructed in Sheffield, England. In the eighteenth century the steel industry was expanding in the United States with considerable production of pig iron and bar iron, with a slitting mill being established in 1793 near Coatesville, Pennsylvania.

In the nineteenth century there was an increase in the size of the mills and their rolled products. This occurred mostly due to the coming of the industrial revolution wherein there was a significant demand for iron and steel. Most of the mills at that time were 2-high. Near the middle of the century in Great Britain a 3-high mill was introduced by R. B. Roden. This was later improved upon by Lauth who used a middle roll with a smaller diameter than the two larger backup rolls. This arrangement offered greater productivity with a significant reduction in the power required. A 4-high arrangement for rolling rails and beams was introduced in 1872 in England and a bit earlier in Germany.

Up until 1890 flat rolled products were considered as plates, sheets, and bars, with certain mills specially constructed to produce some very thin flat products. In 1890 sheets of about 371 mm wide and about 3.56 mm thick were rolled in a mill at Warren, Ohio, and later in 1892 a tandem mill that also rolled sheets was built in Austria. In 1893 a mill was built in Bridgeport, Connecticut that produced thin, narrow sheets of up to 178 mm in width. In these initial applications it was determined that there was a limit to the ratio of width to thickness, between 100 and 160, that could be successfully rolled. Later this was improved upon when a mill at Weirton, West Virginia successfully rolled products of widths and thickness in a ratio of about 220 and in 1923 wider and thinner hot strip of about 914 mm by 1.65 mm was produced on a mill in Ashland Kentucky, which was the beginning of the rolling of much wider strip. A mill at Butler, Pennsylvania in 1926 successfully integrated several methods in their rolling practice that are considered as the first of certain techniques that were applied in modern hot strip mills. In the 1930s there were major advancements in the maximum widths of tandem hot rolled strips from about 889 mm in 1935 to as much as about 2.29 m in the 1940s and thereafter into the 1960s.

The mid-1960s saw a significant improvement in the production capability of the overall hot rolling process, of which tandem hot mills were a part, of up to 1.5–3.5 million tons per year of hot rolled strip. The 1970s saw an increase in even tighter dimensional tolerances and higher rolling speeds, with entire mills fully under computer control. The control since this period also became much more sophisticated with more advanced mathematical models that improved the overall quality of the output. Other modern advances were the replacement of motorized screwdowns with hydraulic cylinders and fully automatic thickness control systems, plus considerable advanced improvement in the overall control of strip thicknesses and tensions throughout the mill that ultimately improved the quality and yield of the final output.

The latter part of the twentieth century and earlier part of the twenty-first century have seen tremendous advancements in the area of tandem hot rolling. While a few of the basic concepts of this area have not changed considerably since the later part of the nineteenth century, the mills have become much larger, faster, and capable of producing products of much larger sizes and closer dimensional tolerances, along with other mills of special design for specific applications (e.g., mini-mills). Almost all of this has resulted from the considerable advancements in various areas of engineering and associated sciences, plus new techniques as discovered in governmental and industrial sponsored research laboratories in academia and industry in several different areas of the world. It is interesting to note, however, that the basic configuration of the 4-high mill has remained as a preferred configuration due to its soundness of design, in spite of the newly developed arrangements such as planetary mills and 6-high mills.

Among the recent advancements in the twenty-first century are several new applications for the endless rolling of hot thin gauge strip. These are mostly based on a process developed by Giovani Arvedi that has been in successful operation since 2009 in Cremona, Italy. In this process hot thin gauge strip is produced in an integrated continuous casting and rolling function which includes groups of tandem hot metal rolling stands. The rolling is endless in that the cutting of the strip during processing, such as prior to its entry to tandem hot rolling, is unnecessary. The overall process produces very high-quality thin strip with a considerable reduction in energy consumption and carbon dioxide emissions when compared to conventional rolling and casting processes. Further detail regarding the earlier and more recent history of tandem hot metal rolling can be found in [1–4].

1.3 Process Overview

The tandem hot metal strip finishing mill is a portion of an overall process that produces coils of metal strip for eventual further processing in a cold rolling mill or other applications. The input to the finishing mill can be in the form of transfer bars that originate as slabs that are discharged from reheating furnaces and then processed to be suitable for entry into the finishing mill. Inputs also can be from other sources such as continuous casting operations. The inputs of transfer bars from reheating furnaces are considered in more detail herein since their characteristics generally impose somewhat more severe requirements on the control of the tandem finishing mill. Thus the concepts developed for a controller that conforms to these requirements would most likely also be generally suitable for other types of inputs to the mill. The overall process wherein the slabs originate from reheating furnaces is depicted in Fig. 1.2.

This process consists of a reheating furnace, a descaler, a roughing mill with multiple stands or a single stand for reversing operations, a coil box, a crop shear, a descaler, a tandem finishing mill generally of 6 or 7 stands, a runout table and one or more coilers for the final product. The functions of these are as follows.

The reheating furnace accepts slabs of metal as produced in previous processes for heating to a temperature suitable for processing in downstream functions, particularly the tandem finishing mill. The slabs are heated to a temperature of about 1065–1315 °C, dependent on the material of the slab, to be suitable for processing downstream. The slabs typically rest on supports that move the slab in a manner to

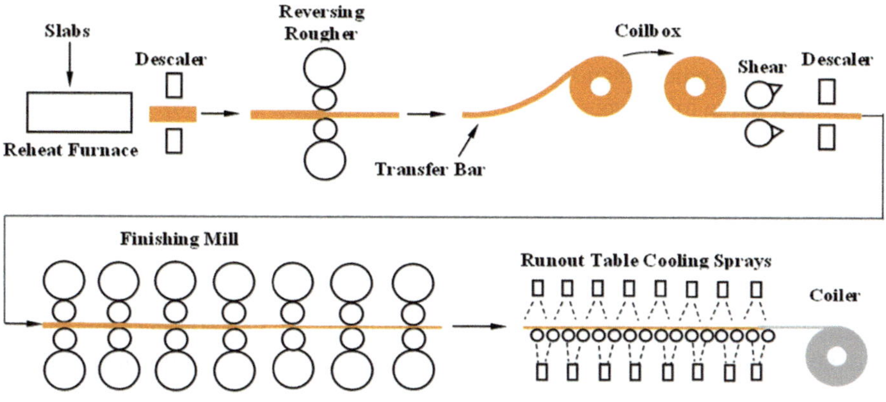

Fig. 1.2 Tandem hot strip finishing mill

reduce colder areas which would result in a higher resistance to deformation during the rolling process. However, there are still cooler areas remaining in the slab that result in disturbances in the temperature and the resistance to deformation of the strip at the entry to the finishing mill that must be handled by the controller during processing of the strip.

The descaler removes scale on the surface of the transfer bar by application of very high pressure water sprays. The removal of scale is necessary to assure that it does not get rolled into the transfer bar in the roughing mill so as to degrade the quality of the rolled metal.

The roughing mill processes the transfer bar to make its dimensions suitable for entry into the finishing mill. In certain mills there are several roughing stands and thus a single pass through the roughing mill is adequate. In other mills, there is a single roughing stand so that the transfer bar is reversed in several passes as needed. While a single roughing stand is less costly, a possible disadvantage is the greater rundown in temperature during the longer time required to process the transfer bar, especially if there is no coil box.

The coil box is a set of rolls that bend the transfer bar into a coil prior to its entry into the finishing mill. By so doing, the reduction in temperature over the length of the bar is reduced, as well as the reduction in the temperature rundown in the entire bar during possible longer periods of time before entering the finishing mill. Thus the temperature of the transfer bar remains high enough for processing in the finishing mill.

The crop shear cuts irregularities in the head and tail of the transfer bar to make it suitable for the entry of the head and the processing of the tail in the finishing mill. The descaler performs a similar function to the descaler at the entry of the roughing mill.

In the tandem finishing mill the transfer bar is converted into strip which, after being processed in the mill, is cooled by water sprays, or water curtains, as it moves over a set of rolls denoted as a runout table. The cooling water is controlled to achieve certain metallurgical properties in the strip as it moves on the runout table, as well as reducing the strip temperature to make it suitable for coiling. The strip is then coiled for use by other functions as needed. More detail on the tandem finishing mill is provided in the following.

1.3.1 The Tandem Hot Strip Finishing Mill

The operation of typical tandem hot strip finishing mills during normal operation is as described in what follows. What is presented is typical, will provide some insight into the overall process and the tandem hot finishing mill process in particular, and can change somewhat for various individual designs and applications.

In this process, the metal strip, in the form of a transfer bar of about 30 mm in thickness, enters the tandem finishing mill typically at a speed of about 1–5 m/s and at a temperature of about 850–1200 °C. In the mill it is passed through a set of six to seven stands each consisting of a pair of independently driven work rolls, each supported by a backup roll of larger diameter. The work rolls and backup rolls are typically 711–762 mm and 1524–1575 mm in diameter, respectively. The distance between adjacent mill stands is on the order of 5.5 m.

When the strip passes through the individual pairs of work rolls the thickness is successively reduced to about 1–12 mm as it exits the mill at a speed of up to 15–20 m/s in certain mills. The reduction in thickness is caused by very high compressive stress in a small region (denoted as the roll gap, or roll bite) between the work rolls. In this region the metal is deformed, and there is slipping between the strip and the work roll surface. The necessary compression force is applied by hydraulic rams or in older installations by a screw arrangement driven by an electric motor. While the strip is exiting the last stand in the mill at a temperature in the range from 750 to 980 °C, its temperature is monitored, and in certain somewhat older mills the mill speed may be increased by a separate system to attempt to control the exit temperature. This increase in speed is denoted as a "zooming" of the mill and is sometimes used wherein there is no coilbox, or if the coilbox is inoperable. A typical 4-high stand is depicted in Fig. 1.3.

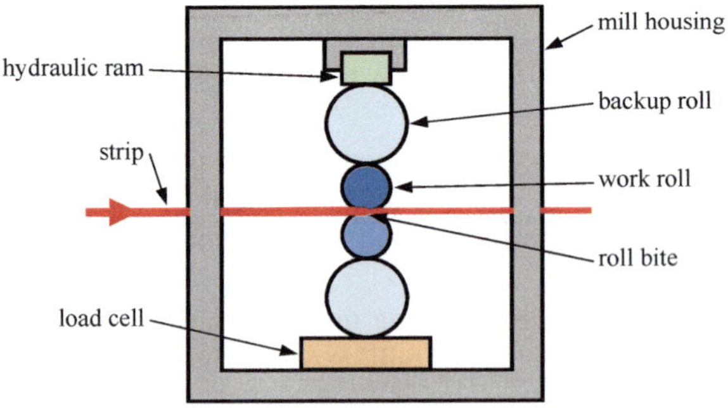

Fig. 1.3 Typical mill stand configuration

1.3.1.1 Loopers

Between each two consecutive mill stands there is a mechanism denoted as a looper. The function of the looper is to assure that the strip between the two consecutive stands can be rolled with little likelihood of tearing or buckling of the strip, which usually results in a wreck (denoted as a cobble) in the mill. Without the looper, the tearing or buckling of the strip would be likely due to the higher temperature of the strip, which reduces its resistance to excessive excursions in tension. It can be noted that while there have been several attempts at looper-less control, so far none have been known to be significantly successful.

The looper consists of an arm and roll driven by hydraulics, to support holding the strip at a reference tension while the looper moves to take up any change in the length of strip between the stands. This change in length is due to differences in the speeds of the strip between the output of an upstream stand and the input of the adjacent downstream stand, which changes are more likely to occur during acceleration, deceleration, and the threading of the mill.

The looper control plays a crucial role in reducing changes in the position of the looper and in excursions in tension in the strip, especially during threading wherein it is desired to maintain nearly constant position and tension as the strip is sequenced from stand to stand. Maintaining nearly constant position and tension, as well as nearly constant thickness, are important to achieving a quality product and support the stability of rolling. A typical schematic for two adjacent stands with a looper driven by hydraulics is depicted in Fig. 1.4.

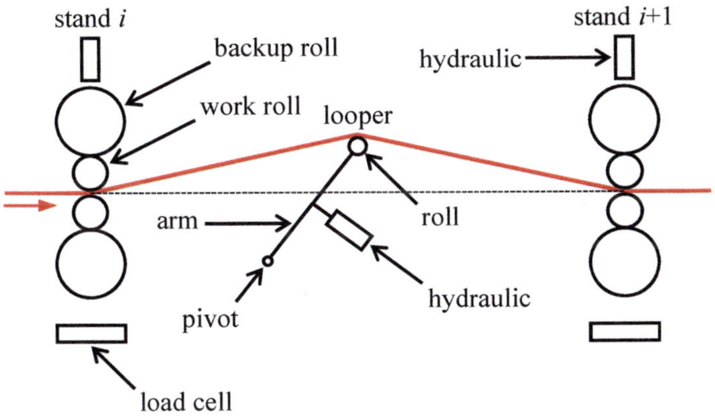

Fig. 1.4 Looper schematic

1.3.1.2 Instrumentation

The lack of space for mounting sensors, a hot rolling environment of high ambient temperatures, plus water sprays for roll coolant are a hostile environment that generally precludes the reliable measurement of certain variables that are important for control. Some examples are the strip speed, thickness, and width between consecutive stands.

However, many other variables can be measured reliably; these include the roll force, looper torque, looper angle and angular velocity, strip tension, roll gap actuator position, and work roll actuator speed. Additionally, more modern techniques, better application of scientific principles, plus experience have resulted in systems that realize reasonably valid measurements of strip thickness and temperature just after the last stand in the mill, and in certain installations the temperature at the entry of the mill.

More recently additional measurements of strip width, the speed of strip at intermediate stands, and some others have been applied in various installations with some success; however, the particular application needs to be carefully considered to assure their overall reliability in a variety of actual operational scenarios.

1.3.1.3 Typical Mill Operating Point Data

Table 1.1 depicts some typical mill operating point data [5].[2] The data are based on actual measurements, except for the temperatures that were estimated based on algorithms used in the mill control computer. In this table the various parameters are the strip thickness (h_{out}) at the stand output (except as shown for the mill entry), strip temperature (T) at the stand entry, the peripheral speed of the work rolls (V_0), and the undeformed radius of the work rolls (R).

Table 1.1 Typical mill operating point data

Stand	h_{out} (mm)	T (°C)	V_0 (m/s)	R (mm)
Mill entry	32.4	–	–	–
Stand 1	21.2	952	1.605	362
Stand 2	13.3	931	2.545	362
Stand 3	9.1	918	3.749	362
Stand 4	6.6	908	5.151	362
Stand 5	4.8	900	7.056	362
Stand 6	3.9	893	8.727	362
Stand 7	3.4	868	10.145	362

[2] It should be noted that the data are for illustrative purposes only, and the actual data depend on the particular design, the products to be processed, operational considerations, plus other considerations peculiar to the installation.

1.3.1.4 Motors and Drives

Most of the modern mills are powered by variable speed AC machines that are synchronous motors or cage rotor induction motors. The drives that power them have the capability for functions that are quite useful for support of plant applications and control. More detail is provided in Chap. 6.

1.4 Subsequent Chapters

Chapter 2 presents a mathematical model of the tandem hot metal strip finishing mill that is useful for control development. Chapter 3 is an overview of the process and its control challenges, Chap. 4 presents conventional control methods that have been successfully applied, while Chap. 5 introduces some advanced techniques for possible improvements in performance. Chapter 6 provides an introduction to the motors and drive systems used in mill applications, so as to provide a fully rounded-out picture of the control of the tandem hot metal strip rolling mill.

References

1. Roberts WL (1983) History of rolling. In: Hot rolling of steel. Marcel Dekker, New York
2. Lenard JG (2014) History of hot strip mills. In: Primer on flat rolling, 2nd edn. Elsevier, Waltham, MA
3. Ginzburg VB, Kaplan NM (2018) Advanced steel plants for hot rolled flat products. Association of Iron and Steel Technology, Warrendale, PA
4. Hohenbichler G et al (2010) Arvedi ESP—technology and plant design. Millennium Steel 82–88
5. Pittner J, Simaan MA (2010) An initial model for control of a tandem hot metal strip rolling process. IEEE Trans Ind Appl 46(1):46–53

Chapter 2
Process Model

2.1 Mathematical Model

A mathematical model of the tandem hot metal rolling process is a set of expressions that mathematically relate the various parameters of the process to each other. Different types of mathematical models of this process have been developed based on their intended use. For example, some of these are models which describe the thermal–mechanical aspects in various areas of the process such as the interactions between the rolls and strip,[1] models which consider the energy aspects during the operation of the process, economic models which are mostly related to cost and profitability, plus others. As used in our work as described herein, the model desired is one that mathematically relates the parameters of the process to each other and also is significant in the development of a control strategy. It is highly desirable that the model represents the physical process to a reasonably close degree, and that this has been confirmed by the use of the model in actual experience in many different applications over several years. Additionally, and highly important, it is that the model must be useful in a practical sense. That is, it must be capable of being easily implemented by users that in most instances have a limited background in advanced mathematical and control theories. Thus the model should be able to be applied in a reasonable manner that is not overly complex and without being computationally demanding, but yet retain the capabilities required for the development and investigation of advanced control techniques.

Generally there are five areas in the development of the model. These are: (1) the prediction of the deformation of the strip, (2) the dynamics of the strip as it is processed from stand to stand during its sequencing through the mill, (3) the prediction of the strip temperature as it moves through the mill, (4) the dynamics of the looper mechanism during its operation and its effects on the looper position and strip tension, and (5) the integration of these four during the various regimes

[1] Herein, and as is often in general usage, the word "strip" will refer to the metal as it is being processed in the mill, and also refer to the mill output.

J. Pittner and M. A. Simaan, *Tandem Hot Metal Rolling Mill Control*, Advances in Industrial Control, https://doi.org/10.1007/978-3-031-58082-6_2

of operation such as threading and fully threaded running, and also considering operation during abnormal scenarios.

In the case of the tandem hot rolling process, the model which follows well represents the aforementioned, as will be noted in the individual descriptions of each part of the model, the associated simulations for model verification as presented herein and by others, and the simulations of overall performance with the controller coupled to the model.

2.2 Theoretical System Equations

In the theoretical system equations of this section, simplified relationships based on certain classical derivations are developed that are relevant to the control of the tandem hot metal rolling process. Included are relationships for specific roll force, thickness of the strip at the entry and exit of a mill stand, interstand tension stress, forward slip, work roll hydraulic cylinder position and control, work roll peripheral speed and control, time for an element of strip to move between adjacent stands, looper angular position, angular velocity, torque, looper hydraulic cylinder control, and others.

In these relationships the following are applicable unless indicated otherwise: The subscript "i" indicates the stand number, with "$i + 1$" indicates the adjacent downstream stand, the subscript "0" indicates the initial value of a variable, and the subscript "op" indicates the desired value of a variable at its operating point.

2.2.1 Specific Roll Force

The estimate of specific roll force is a significant portion of the overall model needed for the development and evaluation of the tandem hot metal rolling controller. In this section a model for the specific roll force is developed using mathematical relationships that are simple but quite useful and have been widely accepted for many years by experienced users in industry and academia and are extensively presented in the literature. Figure 2.1 is a schematic derived from [1] which approximates the roll bite area in which the roll force is applied to reduce the strip thickness as the strip moves through a mill stand.

As shown in this figure the incoming strip is of thickness h_{in} at the strip centerline and is under a tension stress of σ_{in} as it moves toward the input to the roll bite. The output thickness of the strip at its centerline is h_{out} with the strip under a tension stress of σ_{out}. During the rolling process the work roll is flattened elastically from its radius R to the radius R_{p}. Additionally, there is slipping between the surface of the work roll and the strip, so that at the entry of the neutral plane the strip speed V_{in} is less than the peripheral speed V_0 of the roll, and at the exit of the neutral plane the strip speed V_{out} is greater than the peripheral speed of the roll. At the neutral plane

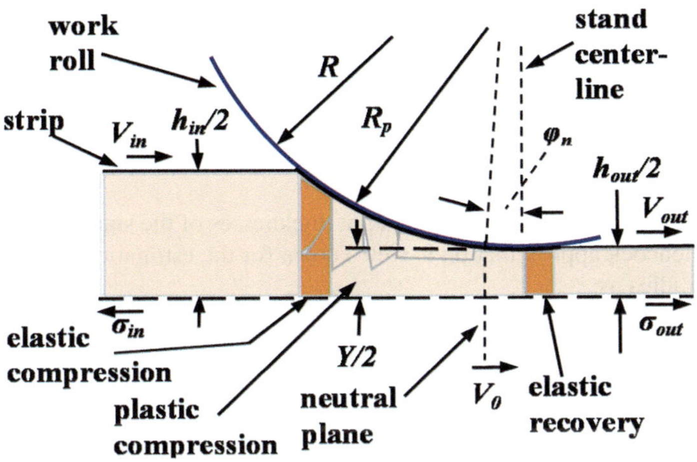

Fig. 2.1 Schematic of the roll bite area[2]

the strip speed is nearly matched to the peripheral speed of the roll. As also depicted in Fig. 2.1, there are three areas of strip deformation: the area of elastic deformation, the area of plastic deformation, and the area of elastic recovery. Other parameters as depicted in this figure are as described more fully in what follows.

The basis for the determination of a relationship for the prediction of specific roll force is Sims' model [3], which is well-known for its usefulness in control development. For our work as described herein, we have enhanced Sims' model by using the empirical results of Shida [4] to describe the resistance to deformation of the material being processed. By using Shida's results, we have improved the estimates of differences in the materials being processed and in the effects of the temperature of the strip.

In Sims' model the specific roll force is expressed as:

$$P = \left(kQ_{\mathrm{p}} - \overline{\sigma}\right)\sqrt{R_{\mathrm{p}}\delta}, \tag{2.1}$$

where P is the specific roll force, k is the resistance to deformation of the material,[3] Q_{p} (2.5) is a factor which compensates for friction and any inhomogeneities of deformation, $\overline{\sigma}$ is the mean tension stress of the strip:

$$\overline{\sigma} = \frac{\sigma_{\mathrm{in}} + \sigma_{\mathrm{out}}}{2}, \tag{2.2}$$

---

[2] Copyright IEEE 2010, Figs. 2.1, 2.3, 2.4, 2.5 and 2.6, and Tables 2.1 and 2.2. Portions of these are reprinted with cosmetic alterations, with permission, from [1]; Unless otherwise noted data depicted are unchanged from [1].

[3] Additional detail describing the empirical results of Shida to estimate k is provided in Appendix.

where σ_{in} and σ_{out} are the strip tension stresses at the stand input and output, R_{p} (2.4) is the deformed work roll radius which is determined using the Hitchcock approximation [5], and δ is the stand draft:

$$\delta = h_{\text{in}} - h_{\text{out}}, \tag{2.3}$$

where h_{in} and h_{out} are the input and output thicknesses of the stand.

The Hitchcock approximation is well-known for the estimation of the deformed work roll radius as:

$$R_{\text{p}} = R\left(1 + \frac{16(1-\nu^2)P}{\pi E\delta}\right), \tag{2.4}$$

where R is the undeformed work roll radius, ν is Poisson's ratio, and E is Young's modulus.

The factor Q_{p} as determined by Sims is:

$$Q_{\text{p}} = \frac{\pi}{2}a\tan^{-1}\sqrt{\frac{r}{1-r}} - \frac{\pi}{4} - a\sqrt{\frac{R_{\text{p}}}{h_{\text{out}}}}\log_{\text{e}}\left(\frac{Y}{h_{\text{out}}}\right) + \frac{1}{2}a\sqrt{\frac{R_{\text{p}}}{h_{\text{out}}}}\log_{\text{e}}\left(\frac{1}{1-r}\right), \tag{2.5}$$

where

$$a = \sqrt{\frac{1-r}{r}}, \tag{2.6}$$

the reduction r is:

$$r = \frac{h_{\text{in}} - h_{\text{out}}}{h_{\text{in}}}, \tag{2.7}$$

and

$$Y = h_{\text{out}} + R_{\text{p}}\phi_{\text{n}}^2. \tag{2.8}$$

As shown in Fig. 2.1 Y is an approximation of the strip thickness at the neutral plane, wherein the neutral plane is a position along the arc of contact of the strip with the work roll, and the position of the neutral plane from the stand center line is noted by the angle ϕ_{n}, where

$$\phi_{\text{n}} = \frac{\tan(\hat{a} + b)}{\sqrt{\frac{R_{\text{p}}}{h_{\text{out}}}}}, \tag{2.9}$$

and where

$$\hat{a} = \frac{\pi}{8} \frac{\log_e(1-r)}{\sqrt{\frac{R_p}{h_{out}}}}, \tag{2.10}$$

and

$$b = \frac{1}{2} \tan^{-1} \sqrt{\frac{r}{1-r}}. \tag{2.11}$$

2.2.2 *Input and Output Strip Thicknesses*

The input strip thickness to the first mill stand $h_{in,1}$ is the input strip thickness to the mill. The input thickness to each of the remaining stands $h_{in,i}$ ($i = 2, 3, 4 \ldots$) is the output thickness $h_{out,i-1}$ as tracked from the output of the adjacent upstream stand using the interstand time delay (2.27).

The output strip thickness of a mill stand is estimated by a linear approximation as:

$$h_{out} = S + S_0 + \frac{F}{M}, \tag{2.12}$$

where S is the position of the actuator for the roll bite position, $(S + S_0)$ is the intercept of the h_{out} axis of the linearized approximation of the mill stretch curve as depicted in Fig. 2.2, $F = PW$ is the total rolling force, where P is the specific rolling force (2.1), W is the strip width, and M is the mill modulus which, under the application of F, is the elastic stretch of the mill.

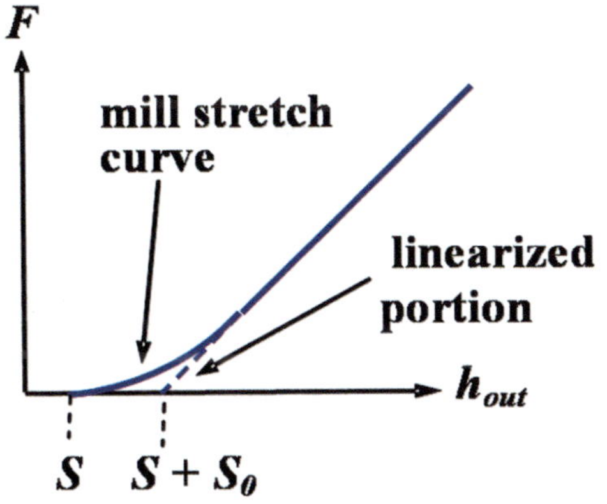

Fig. 2.2 Output strip thickness approximation

This approximation (2.12) is often denoted as a “gaugemeter” estimate, or a BISRA estimate, which was developed by the British Iron and Steel Association in the 1950s. More detail regarding the usage of this estimate is described in Chap. 3.

2.2.3 Forward Slip

The forward slip is a means of estimating the strip speed at the output of a mill stand just after the roll bite. It is expressed as the ratio of the increase in the strip speed just after the roll bite in relation to the strip speed at the neutral plane:

$$f = \frac{V_{\text{out}} - V_0}{V_0}, \tag{2.13}$$

where V_{out} and V_0 are as previously noted in Sect. 2.2.1. In Sims' work the forward slip is developed as:

$$f = \left(\frac{\delta}{h_{\text{out}}}\right)\left(\frac{\phi_{\text{n}}}{\phi_1}\right)^2, \tag{2.14}$$

where

$$\phi_1 = \left(\frac{\delta}{R_{\text{p}}}\right)^{\frac{1}{2}}, \tag{2.15}$$

and other parameters are as noted previously. In the work described herein we have used a model that is more useful for control development as it explicitly includes the effects of the strip tension. This alternative model was developed by Ford for cold mill applications [6], but has been successfully applied to hot mills by using the friction coefficient μ to represent sticking friction for hot mills rather than sliding friction as in the case of cold mills.[4]

In Ford's model the forward slip is expressed as:

$$f = \left(\frac{R_{\text{p}}}{h_{\text{out}}}\right)(\beta_{\text{n}})^2, \tag{2.16}$$

where

$$\beta_{\text{n}} = \frac{\phi_1}{2} - \frac{\delta k + \sigma_{\text{in}} h_{\text{in}} - \sigma_{\text{out}} h_{\text{out}}}{4k R_{\text{p}} \mu}, \tag{2.17}$$

and where the friction coefficient is approximated as in Roberts [7] as:

$$\mu = 0.00027 T_{\text{F}} - 0.08, \tag{2.18}$$

[4] The verification of this substitution for use in the hot mill is addressed more fully in Sect. 2.4.3.

where μ is the friction coefficient, T_F is the strip temperature in degrees Fahrenheit, and other parameters are as previously noted.

2.2.4 Work Roll Actuator Position

The closed loop position controller for the hydraulic cylinder (i.e., the actuator) that moves the work roll and sets its position at the roll bite is modeled as a first-order lag:

$$\frac{\mathrm{d}S}{\mathrm{d}t} = \frac{U_S}{\tau_S} - \frac{S}{\tau_S}, \quad S(0) = S_0, \tag{2.19}$$

where S is the cylinder position, U_S is the position reference, and τ_S is the closed loop time constant.

2.2.5 Work Roll Peripheral Speed

The closed loop controller for the peripheral speed of the work roll at the neutral plane is modeled as a first-order lag:

$$\frac{\mathrm{d}V}{\mathrm{d}t} = \frac{U_V}{\tau_V} - \frac{V}{\tau_V}, \quad V(0) = V_0, \tag{2.20}$$

where V is the work roll speed, U_V is the speed reference, and τ_V is the closed loop time constant.

2.2.6 Looper Characteristics

The relationships describing the looper and its controller are as follows[5]:

The angle of the looper position is expressed as:

$$\frac{\mathrm{d}\theta}{\mathrm{d}t} = \omega, \quad \theta(0) = \theta_0, \tag{2.21}$$

where θ is the angle of the looper arm, and ω is the angular velocity of the looper as determined using Newton's second law of motion:

$$\frac{\mathrm{d}\omega}{\mathrm{d}t} = \frac{1}{J_{\mathrm{lpr}}}\left[M_{\mathrm{lpr}} + M_{\mathrm{fct}} + M_{\mathrm{ld}}\right], \quad \omega(0) = 0, \tag{2.22}$$

[5] Additional details regarding the looper characteristics and typical values of various parameters during actual operation are provided in Appendix.

where J_{lpr} is the moment of inertia of the looper.

M_{lpr} is the torque applied to the looper by a hydraulic cylinder under closed loop control, where M_{lpr} is approximated as a single first-order lag, $U_{\text{M}_{\text{lpr}}}$ is the torque reference, and τ_{M} is the closed loop time constant:

$$\frac{\mathrm{d}M_{\text{lpr}}}{\mathrm{d}t} = \frac{U_{\text{M}_{\text{lpr}}}}{\tau_{\text{M}}} - \frac{M_{\text{lpr}}}{\tau_{\text{M}}}, \quad M_{\text{lpr}}(0) = M_{\text{lpr},0}. \tag{2.23}$$

M_{fct} is the frictional torque of the looper, which is approximated to be viscous friction, where

$$M_{\text{fct}} = k_{\text{vis}}\omega, \tag{2.24}$$

and where k_{vis} is the viscous friction constant.

The looper load torque M_{ld} is determined by the torques resulting from the strip tension force M_{ten}, the strip weight M_{swt}, the looper mass M_{lmas}, and the strip bending M_{bnd} at the looper roll, as:

$$M_{\text{ld}} = M_{\text{ten}} + M_{\text{swt}} + M_{\text{lmas}} + M_{\text{bnd}}. \tag{2.25}$$

2.2.7 *Interstand Strip Tension Stress*

The interstand strip tension stress σ is modeled based on a relationship for Young's modulus wherein the strain depends on the length of the strip between adjacent mill stands:

$$\frac{\mathrm{d}\sigma}{\mathrm{d}t} \equiv \dot{\sigma} = \frac{E}{L_0}\left[\frac{\mathrm{d}L(\theta(t))}{\mathrm{d}t} + V_{\text{in},i+1} - V_{\text{out},i}\right], \quad \sigma(0) = \sigma_0, \tag{2.26}$$

where E is Young's modulus, L_0 is the distance between the centerlines of adjacent stands, $L(\theta(t))$ is the strip length between adjacent stands which depends on the position of the looper, $V_{\text{in},i+1}$ is the strip speed at the input to stand $i+1$, and $V_{\text{out},i}$ is the strip speed at the output of stand i.

2.2.8 *Interstand Time Delay*

The interstand time delay is approximated as:

$$\tau_{\text{d},i,i+1} = \frac{L}{V_{\text{out},i}}, \tag{2.27}$$

where L is the length of strip between stands i and $i+1$ as in (2.26).

2.3 State-Space Equations

The previous relationships describing the tandem hot metal rolling process are put into state-space form for use in the development and evaluation of the controller. Specifically:

$$\dot{x} = A(x)x + Bu, \quad x(0) = x_0, \tag{2.28}$$

and

$$y = C(x)x, \tag{2.29}$$

where $x \in R^n$ $(n = 38)$ is a vector whose elements represent the individual state variables, $A(x) \in R^{n\times n}$ and $C(x) \in R^{p\times n}$ $(p = 26)$ are state-dependent matrices, $u \in R^m$ $(m = 20)$ is a vector whose elements represent the individual control variables, $y \in R^p$ is a vector whose elements represent the individual output variables, and $B \in R^{n\times m}$ is a constant matrix. The variables represented by each of the elements of the state, output, and control vectors are as noted in Table 2.1 [1]. The elements of the $A(x)$, $C(x)$, and B matrices are as given in Appendix.

Table 2.1 Variable assignments for elements of the state, control, and output vectors

State variables		Control variables	Output variables	
x_1 (σ_{12})	x_{21} (M_{12})	u_1 (U_{S1})	y_1 (h_{out1})	y_{14} (P_1)
x_2 (σ_{23})	x_{22} (M_{23})	u_2 (U_{S2})	y_2 (h_{out2})	y_{15} (P_2)
x_3 (σ_{34})	x_{23} (M_{34})	u_3 (U_{S3})	y_3 (h_{out3})	y_{16} (P_3)
x_4 (σ_{45})	x_{24} (M_{45})	u_4 (U_{S4})	y_4 (h_{out4})	y_{17} (P_4)
x_5 (σ_{56})	x_{25} (M_{56})	u_5 (U_{S5})	y_5 (h_{out5})	y_{18} (P_5)
x_6 (σ_{67})	x_{26} (M_{67})	u_6 (U_{S6})	y_6 (h_{out6})	y_{19} (P_6)
x_7 (S_1)	x_{27} (θ_{12})	u_7 (U_{S7})	y_7 (h_{out7})	y_{20} (P_7)
x_8 (S_2)	x_{28} (θ_{23})	u_8 (U_{V1})	y_8 (σ_{12})	y_{21} (θ_{12})
x_9 (S_3)	x_{29} (θ_{34})	u_9 (U_{V2})	y_9 (σ_{23})	y_{22} (θ_{23})
x_{10} (S_4)	x_{30} (θ_{45})	u_{10} (U_{V3})	y_{10} (σ_{34})	y_{23} (θ_{34})
x_{11} (S_5)	x_{31} (θ_{56})	u_{11} (U_{V4})	y_{11} (σ_{45})	y_{24} (θ_{45})
x_{12} (S_6)	x_{32} (θ_{67})	u_{12} (U_{V5})	y_{12} (σ_{56})	y_{25} (θ_{56})
x_{13} (S_7)	x_{33} (ω_{12})	u_{13} (U_{V6})	y_{13} (σ_{67})	y_{26} (θ_{67})
x_{14} (V_1)	x_{34} (ω_{23})	u_{14} (U_{V7})		
x_{15} (V_2)	x_{35} (ω_{34})	u_{15} (U_{MIpr12})		
x_{16} (V_3)	x_{36} (ω_{45})	u_{16} (U_{MIpr23})		
x_{17} (V_4)	x_{37} (ω_{56})	u_{17} (U_{MIpr34})		
x_{18} (V_5)	x_{38} (ω_{67})	u_{18} (U_{MIpr45})		
x_{19} (V_6)		u_{19} (U_{MIpr56})		
x_{20} (V_7)		u_{20} (U_{MIpr67})		

2.4 Model Validation

An acceptable validation of the model should be based on valid data obtained from actual operations, and be consistent with models developed by others that were in use for several years and suitable for supporting the development and investigation of control. However, in the case of the tandem hot metal rolling mill, while some data can be obtained, getting considerable valid and useful data from actual operating installations is generally quite difficult and limited in most cases. This is mostly due to the dearth of certain reliable measurements that are precluded by the harsh environment of the process, the difficulty of interrupting actual mill production to verify certain model characteristics, and the significant expense required of the mill owner.

Considering this, we have validated the previous roll force model (2.1) by comparing (1) against a thermal–mechanical finite element model (FEM) by others [8] that have considerable background and recognized expertise in this area, (2) against the data [8] that could be obtained in support of this FEM analysis, and (3) against data obtained from another operating mill [9]. Additionally, in the case of the looper, verification was by comparison with data obtained by others [10] that have similar experience in support of looper modeling. For the validation, we have considered the roll force model, and the looper model as being most significant, and a model of the forward slip as also being useful.

It should be noted that it is not possible to successfully operate the mill open loop. This is due to the uncertainties present in any mill-model relationship, so that during operation the actual mill cannot ideally conform to the model used for the setup of the mill controller. Additionally, the model cannot predict disturbances that will occur during actual operation. Thus, while the aforementioned is generally suitable for verification of steady-state conditions, for a measure of validation of dynamic responses of the mill under open loop conditions, we have applied by simulation small step changes in various variables of the model, open loop. Exceptions are the closed loop controllers for work roll speeds, work roll gap cylinder positions, and looper torques. In these cases, each of these controllers is simulated by a closed loop single first-order lag, to approximately simulate a nearly fixed condition of these functions under perturbations in certain areas in the remainder of the mill. As it is nearly impossible to obtain data from actual operation under similar conditions, we have considered the response to a particular perturbation reasonable if it is well behaved and not excessively severe or extreme.

2.4.1 Roll Force Model

The results of the roll force model used herein (2.1) are compared to the results of the roll force FEM of [8]. In this comparison, the following in the model of (2.1) were

Table 2.2 Data for the mill associated with the FEM

Pass	h_{out} (mm)	Work roll		
		RPM	Radius (mm)	V_0 (m/min)[8]
0	38.8	–	–	–
1	21.6	32	360	72
2	14.4	52	336	110
3	8.6	80	353	177
4	6.1	116	343	250
5	4.7	155	388	378
6	3.9	184	348	402
7	3.5	204	369	473

compared to those of the FEM in steady state: the strip temperatures[6] and strip input thicknesses at each stand, thickness reductions at each stand, work roll diameters, and the carbon equivalent of the strip being processed. An additional comparison was done, except with the measured mill temperatures as given in [9], wherein the measured temperatures were taken to represent the actual temperatures.[7] During the validation the strip tension stresses were taken to be at a typical nominal value of 0.01 kN/mm^2 since in the FEM analysis results the tension that was used was not specified. This was then compared with zero tension with only minor differences in the results. The results of the simulations using the FEM model are depicted in Table 2.3, wherein the roll forces estimated using the model of (2.1) reasonably follow what is estimated using the FEM. The strip output thickness, work roll RPM, radius, and peripheral speed for each mill stand associated with the FEM are given in Table 2.2.

Additionally, data from a second mill [9] in actual operation were examined for comparison with the specific roll force model (2.1). As in the previous case, the model was set to use the strip input thicknesses at each stand, thickness reductions at each stand, work roll diameters, and the carbon equivalent taken as typical of the strip being processed, based on actual measurements and production schedules. The temperatures at the entry of a mill stand were determined by a separate offline system whose algorithm for temperature estimation was unknown. The results of the simulation are presented in Table 2.5. As can be seen in Table 2.5, there are greater discrepancies between the model and the measured roll force, than in the previous case. It is felt that this might be due to uncertainties in the unknown algorithm that was used to estimate the strip temperatures. These temperatures were each adjusted

[6] The FEM temperatures were preferred as it was considered that, due the harsher mill environment, they may have slightly less uncertainty than the measured temperatures.

[7] A detailed analytical estimation of specific mill temperatures is considered to be outside the intended scope of this effort. Except as noted otherwise, the strip temperatures used are taken to be those determined in the FEM analysis, with some approximations as noted in [8].

[8] The speed V_0 is as noted in Sect. 2.2.1.

Table 2.3 Comparison of the model of the specific roll force (2.1) using the FEM temperatures T_{FEM}, with the specific roll force using the model and the measured temperatures T_{Mill}, and the specific roll force using the FEM analysis, with the reduction r (2.7)[9]

Pass	h_{out} (mm)	r	T_{FEM} (°C)	T_{Mill} (°C)	Specific roll force (kN/mm)		
					Model (T_{FEM})	Model (T_{Mill})	FEM (T_{FEM})
0	38.8	–	1058	1058	–	–	–
1	21.6	0.44	988	993	15.6	15.5	16.4
2	14.4	0.33	973	986	12.6	12.2	12.3
3	8.6	0.40	957	968	16.7	16.5	14.6
4	6.1	0.29	938	951	12.5	12.0	10.5
5	4.7	0.23	922	935	11.8	11.3	9.4
6	3.9	0.17	904	913	8.1	7.8	6.4
7	3.5	0.10	894	898	4.3	4.2	3.7

Table 2.4 Data for the second mill

Pass	h_{out} (mm)	Work roll		
		RPM	Radius (mm)	V_0 (m/min)[10]
0	32.4	–	–	–
1	21.2	42	362	96
2	13.3	67	362	152
3	9.1	99	362	223
4	6.6	136	362	309
5	4.8	186	362	423
6	3.9	230	362	523
7	3.4	268	362	610

to about 5 or 10 percent which improved the correlation with the model, so that there was general overall agreement with the model. The strip output thickness, work roll RPM, radius, and peripheral speed for each mill stand are given in Table 2.4.

---

[9] Copyright IEEE 2010, portions of Tables 2.3, 2.4, 2.5 and 2.7, are reprinted with cosmetic alterations, with permission, from [2]. Unless otherwise noted, data depicted are unchanged from [2].

[10] The speed V_0 is as noted in Sect. 2.2.1.

Table 2.5 Comparison of the model of the specific roll force (2.1) using the estimated temperatures T_{Mill} of the second mill, with the specific roll force using the mill data and the temperatures T_{Mill}

Pass	h_{out} (mm)	r	T_{Mill} (°C)	Specific roll force (kN/mm)	
				Model (T_{Mill})	Mill (T_{Mill})
0	32.4	–	–	–	–
1	21.2	0.35	952	13.9	11.0
2	13.3	0.37	931	16.3	10.7
3	9.1	0.32	918	14.7	9.8
4	6.6	0.27	908	13.1	9.9
5	4.8	0.27	900	15.7	8.8
6	3.9	0.19	893	9.5	7.3
7	3.4	0.14	868	5.8	5.4

2.4.2 Looper Model

The looper model as described in Sect. 2.2.6 and Appendix was evaluated by comparison with data derived from loopers at several installations. The data were collected by others [10] with considerable experience in this area. The looper selected for comparison was between stands 6 and 7, which is typical for other loopers in the mill. With a tension stress of 0.01 kN/mm^2, looper torques for tension, strip weight, looper mass, strip bending, and the total torque were compared against each point of the looper position as noted in Table 2.6. A summary of the results of the simulations is given in Table 2.6 for steady-state operation. As noted in this table, the model described herein conforms well to the data derived in [10].

Table 2.6 Comparison of the model of the looper (Sect. 2.2.6) with data obtained from actual loopers

	Looper	
	Model	Data
Position in mill	Stands 6, 7	Stands 6, 7
Operating point		
Strip width (mm)	1290	1290
Strip thickness (mm)	3.9	3.9
Tension stress (kN/mm^2)	0.01	0.01
Positions simulated	7.5–22.5° at 2.5° increments	7.5–22.5° at 2.5° increments
Mean deviation from data (all points)	2.62%	–
Standard deviation from data (all points)	2.07%	–

2.4.3 Forward Slip

The validation of the model (2.16) as used for estimation of the forward slip was by comparison with Sims' calculations (2.14) for the first 5 stands of a typical mill.[11] Table 2.7 depicts the results of the comparison which shows reasonable conformance to Sims' calculation.

2.4.4 Dynamic Responses

As previously noted, it is nearly impossible to fully validate dynamic characteristics of the mill model under open loop conditions or during actual operations. However, a measure of open loop mill behavior can be seen by the application of small perturbations in certain mill areas. Under these conditions, the response to a certain perturbation is considered reasonable if it is well behaved and not excessively severe or extreme. On this basis we have applied small steps in certain variables of the model, open loop, that are kept constant unless they are perturbed. These are the incoming strip thickness, strip temperature, and material, plus the references for controllers that are simulated as closed loop first-order lags. These variables are the references for the controllers for work roll speeds, work roll gap cylinder positions, and looper torques.

The small step changes were $\pm$ 0.1% in each of the references for work roll speed and roll gap cylinder position controllers; $\pm$ 1.0% for the incoming strip thickness and in each reference for the looper torque controllers; $\pm$ 0.3% in the incoming strip temperature; and $\pm$ 5.0% in the incoming strip equivalent carbon content. At the application of each perturbation the following responses were taken: strip temperatures at each stand, plus stand output speeds and strip thicknesses; strip tensions, roll forces, forward slips, and looper positions. In each case the responses were not extreme and reasonably well behaved. Some examples (Figs. 2.3, 2.4, 2.5 and 2.6) are given in [1], with some further evaluation in Appendix.

Table 2.7 Comparison of the model of forward slip (2.16) with Sims' calculations (2.14)

Stand	Forward slip (%)	
	Model	Sims'
1	11.6	10.3
2	8.2	8.1
3	11.8	10.5
4	8.1	7.6
5	6.4	6.0

[11] In stands 6 and 7 uncertainties in the forward slip generally are of lesser concern since typically the forward slips in these stands are much less than unity.

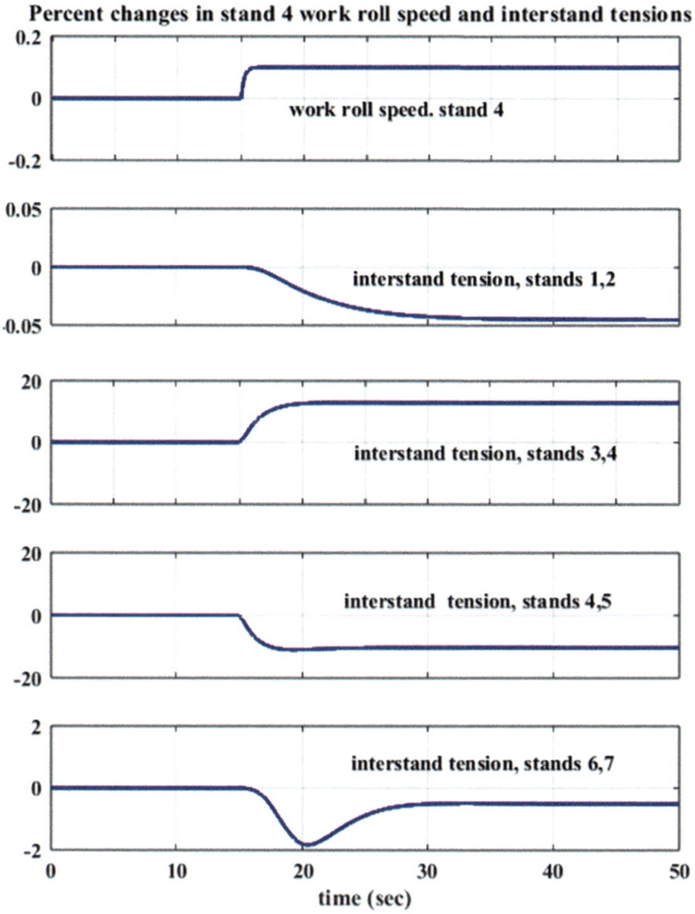

Fig. 2.3 Percent changes in stand 4 work roll speed and interstand tensions for a 0.1% step change in stand 4 work roll speed ref applied at 15 s

It can be noted that the application of small perturbations at a single mill stand can often contribute to a better understanding of the dynamics throughout the entire mill (open loop), so that the responses throughout the mill can be useful in the development and design of mill control techniques. As an example, in Fig. 2.3, a small perturbation in the work roll speed of stand 4 would cause significant changes in the tensions between stands 3 and 4, and between stands 4 and 5, but also can propagate downstream to be seen in the tensions between stands 6 and 7, and upstream (but with a lesser impact) on the tensions between stands 1 and 2. Thus in most cases the control must acceptably handle not only a single mill area but also the effects on other significant mill areas. This is addressed more fully in Appendix and Chaps. 4 and 5.

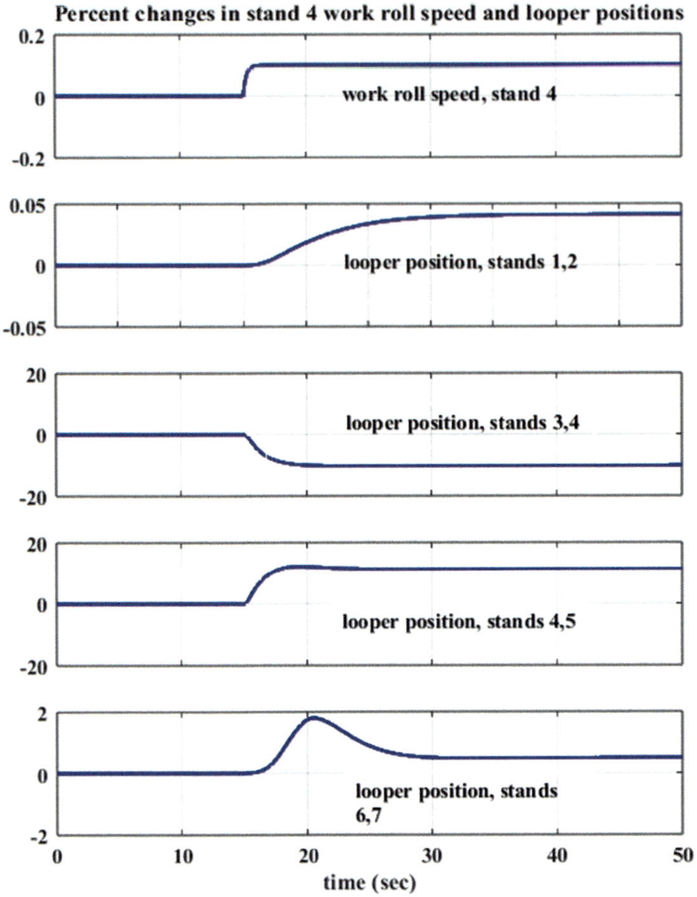

Fig. 2.4 Percent changes in stand 4 work roll speed and looper positions for a 0.1% step change in stand 4 work roll speed ref applied at 15 s

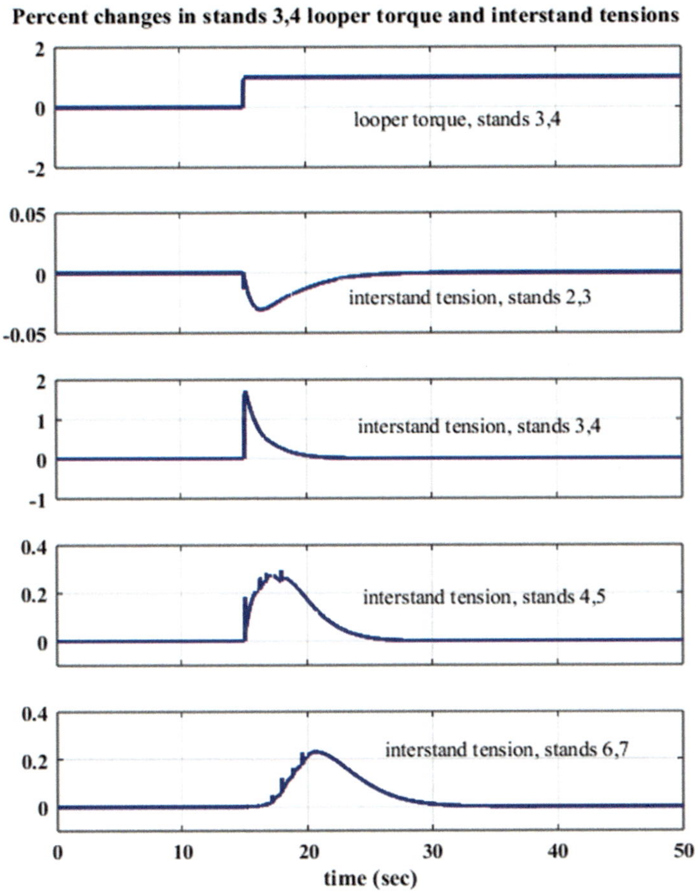

Fig. 2.5 Percent changes in looper torque and interstand tensions for a 1% step change stands 3, 4 looper torque ref applied at 15 s

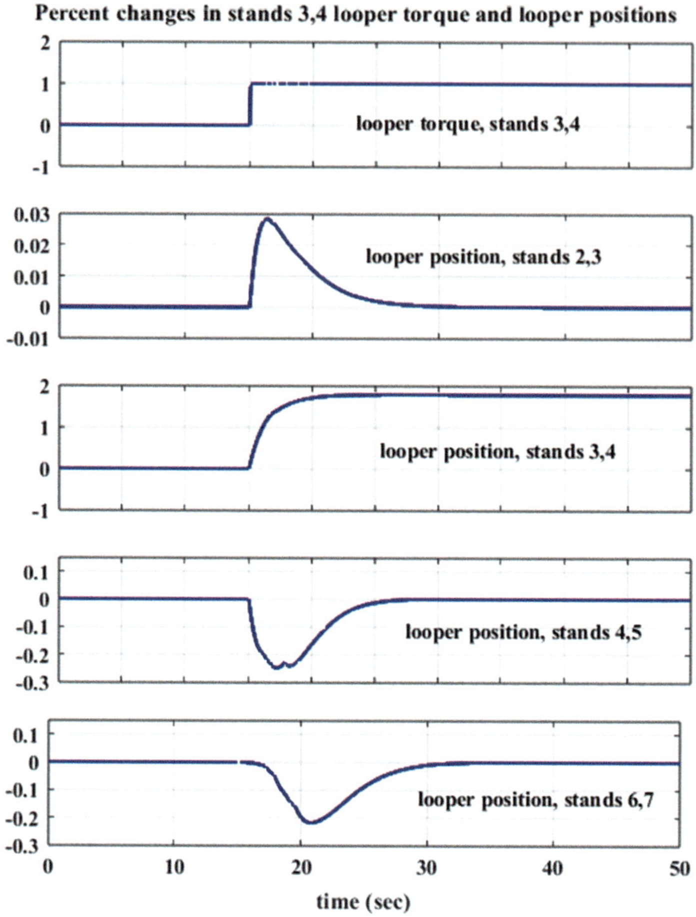

Fig. 2.6 Percent changes in stands 3, 4 looper torque and looper positions for a 1% step change in stands 3, 4 looper torque ref applied at 15 s

Appendix

The sections of this Appendix provide material that supplements what is presented in this chapter. Provided are some details regarding the results of Shida's work as applied to the estimate of the specific roll force, detail of the characteristics of the looper, detail of the elements of the state-space matrices, and an example of how the plots of responses to small perturbations can help in better understanding the associated models.

Shida's Relationships

For the parameter k (resistance to deformation) as used in (2.1) for specific roll force we have applied the comprehensive set of the empirical results of Shida [4]. These results were developed by his extensive testing of carbon steel of various temperatures and materials. The validity and usefulness of these results have been recognized by others with extensive expertise in this area. These expressions are closed-form estimates of the temperatures and material characteristics of the strip and thus are quite useful for the development and evaluation of the controller as they simplify the method of modeling, and compare reasonably well to what is presented in the several plots in Sims' work. The expressions provided are estimates of k in terms of the temperature in degrees Celsius, the carbon content of the material being processed, the strain, and the strain rate.[12] The following are applicable in these expressions: temperature T (700–1200 °C); carbon content C ($< 1.2\%$ (weight)); strain ε ($< 70\%$); the strain rate $\dot{\varepsilon}$ (0.1–100 s^{-1}).

Shida's relation for the resistance to deformation is:

$$k = k_f f \left(\frac{\dot{\varepsilon}}{10} \right)^m, \tag{2.30}$$

where if

$$\overline{T} \geq 0.95 \frac{C + 0.41}{C + 0.32}, \tag{2.31}$$

where

$$\overline{T} = \frac{T + 273}{1000}, \tag{2.32}$$

then

$$k_f = 0.28 \exp\left(\frac{5}{\overline{T}} - \frac{0.01}{C + 0.05} \right), \tag{2.33}$$

and

$$m = (-0.019C + 0.126)\overline{T} + (0.075C - 0.05). \tag{2.34}$$

However if

$$\overline{T} < 0.95 \frac{C + 0.41}{C + 0.32}, \tag{2.35}$$

[12] In these expressions it should be noted that the characteristics presented are mathematical descriptions of Shida's empirical results and do not represent derivations based on actual models of the physical entities being considered.

then

$$k_{\mathrm{f}} = 0.28q\left(C, \overline{T}\right) \exp\left(\frac{C + 0.32}{0.19(C + 0.41)} - \frac{0.01}{C + 0.05}\right), \tag{2.36}$$

where

$$q\left(C, \overline{T}\right) = 30(C + 0.9)\left(\overline{T} - 0.95\frac{C + 0.49}{C + 0.42}\right)^2 + \frac{C + 0.06}{C + 0.09}, \tag{2.37}$$

and

$$m = (0.081C - 0.154)\overline{T} - 0.019C + 0.207 + \frac{0.027}{C + 0.32}. \tag{2.38}$$

The parameter f in (2.30) is:

$$f = 1.3(5\varepsilon)^n - 1.5\varepsilon, \tag{2.39}$$

where

$$n = 0.41 - 0.07C. \tag{2.40}$$

The (mean) strain rate used in (2.30) is:

$$\dot{\varepsilon} = \frac{2\pi N}{60\sqrt{r}}\sqrt{\frac{R_{\mathrm{p}}}{h_{\mathrm{in}}}\log_{\mathrm{e}}\left(\frac{1}{1 - r}\right)}, \tag{2.41}$$

where N is the work roll revolutions/minute, and the strain is:

$$\varepsilon = \log_{\mathrm{e}}\left(\frac{1}{1 - r}\right), \tag{2.42}$$

and R_{p}, h_{in}, and r are as in (2.3), (2.4), and (2.7).

Looper Characteristics

The looper characteristics as denoted in what follows are based on data available from an actual installation and as described in [10]. A schematic of this typical looper between mill stand 6 and stand 7, and similar for all loopers in the rest of the mill, is depicted in Fig. 2.7.

Dimensions: $L_0 = 5.478$ m, $y = 0.191$ m, $a = 1.943$ m, $r = 0.152$ m, $l = 0.762$ m. Calculation of $l1$ and $l2$: Taking $A = l \sin\theta$, $B = l \cos\theta$, $C = A + r$, $D = B + a$,

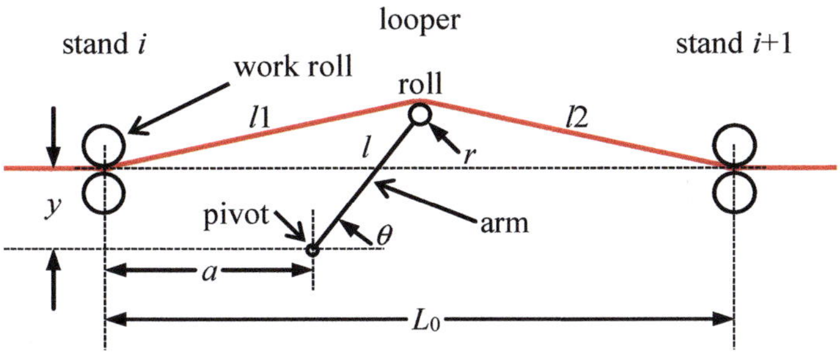

Fig. 2.7 Looper schematic

$$l1 = \left((r - y + A)^2 + (a + B)^2\right)^{\frac{1}{2}} = 2.684\,\mathrm{m}, \tag{2.43}$$

$$l2 = \left((r - y + A)^2 + (L_0 - a - B)^2\right)^{\frac{1}{2}} = 2.806\,\mathrm{m}. \tag{2.44}$$

Operating range: Maximum $\theta = 40°$, passline $\theta = 2.9°$.

Operating point, steady state: $\theta = 15°$, $\omega = 0$ rad/s, $l1$ and $l2$ as above.

Moment of inertia: Mass of looper arm: 300 kg, mass of looper roll: 500 kg,

Calculation of moment of inertia:

$$J_{\mathrm{lpr}} = \frac{300l^2}{3} + 500\left(l^2 + \frac{r^2}{2}\right) = 0.3542\left(\frac{\mathrm{kg\,m^2}}{1000}\right). \tag{2.45}$$

Viscous friction coefficient:

$$k_{\mathrm{vis}} = -2.0\,\mathrm{kN\,m/rad/s}, \text{ as in (2.24)}. \tag{2.46}$$

Looper and load torques:

Calculation of strip tension torque:

$$M_{\mathrm{ten}} = -F_{67}\left[C\left(\frac{D}{l1} - \frac{(L_0 - D)}{l2}\right) - B(C - y)\left(\frac{1}{l1} + \frac{1}{l2}\right)\right]$$
$$= -4.289\,\mathrm{kN\,m}, \tag{2.47}$$

where F_{67} is the tension force $= 50.310$ kN.

Calculation of strip weight torque:

$$M_{\mathrm{swt}} = -(l1 + l2)Wh_{\mathrm{out6}}B\left(74.556 \times 10^{-6}\right) = -1.515\,\mathrm{kN\,m}, \tag{2.48}$$

where $W = 1290$ mm is the strip width, and $h_{\text{out6}} = 3.9$ mm is the strip thickness.

Calculation of looper mass torque:

$$M_{\text{lmas}} = -B(0.00981)\left(\frac{300}{2} + 500\right) = -4.693\,\text{kN m}. \tag{2.49}$$

Calculation of strip bending torque:

$$M_{\text{bnd}} = -\left(\frac{5 \times 10^{-5}}{4}\right) B W h_{\text{out6}}^2 \left(\frac{1}{l1} + \frac{1}{l2}\right) = -0.132\,\text{kN m}. \tag{2.50}$$

Calculation of load torque:

$$M_{\text{ld}} = M_{\text{ten}} + M_{\text{swt}} + M_{\text{lmas}} + M_{\text{bnd}} = -10.629\,\text{kN m}. \tag{2.51}$$

Calculation of looper torque at steady state:

$$M_{\text{lpr}} = -M_{\text{ld}} = 10.629\,\text{kN m}. \tag{2.52}$$

Calculation of d$L(\theta(t))$/dt:

$$\frac{\mathrm{d}L(\theta(t))}{\mathrm{d}t} = \frac{\mathrm{d}L}{\mathrm{d}\theta}\frac{\mathrm{d}\theta}{\mathrm{d}t}, \tag{2.53}$$

where

$$L = l1 + l2, \tag{2.54}$$

$$\frac{\mathrm{d}L}{\mathrm{d}\theta} = \frac{\mathrm{d}(l1)}{\mathrm{d}\theta} + \frac{\mathrm{d}(l2)}{\mathrm{d}\theta}, \tag{2.55}$$

$$\frac{\mathrm{d}(l1)}{\mathrm{d}\theta} = \frac{l((r-y)\cos\theta - a\sin\theta)}{\sqrt{l1}}, \tag{2.56}$$

$$\frac{\mathrm{d}(l2)}{\mathrm{d}\theta} = \frac{l((r-y)\cos\theta - (L_0 - a)\sin\theta)}{\sqrt{l2}}, \tag{2.57}$$

and where

$$\frac{\mathrm{d}\theta}{\mathrm{d}t} = \omega,\;\; L(\theta(0)) = l1(\theta(0)) + l2(\theta(0)),\;\; \text{and } \theta(0) = \theta_0. \tag{2.58}$$

State-Space Equations

Elements of $A(x)$:

$A(x)$ is a zero matrix except for the following elements:

$$A_{1,14} = \frac{E}{L_0}\left(- V_{\text{out1}} - k_{14}\frac{\mathrm{d}L}{\mathrm{d}t}_{12}\right)\left(\frac{1}{x_{14}}\right), \tag{2.59}$$

$$A_{1,15} = \frac{E}{L_0}\left(V_{\text{out2}}\frac{h_{\text{out2}}}{h_{\text{in2}}} + (k_{14}+1)\frac{\mathrm{d}L}{\mathrm{d}t}_{12}\right)\left(\frac{1}{x_{15}}\right), \tag{2.60}$$

wherein for example k_{14} is determined as 9999 during our initial work, and may be adjusted for actual usage depending on the hardware and software characteristics of the particular installation, $L = l1 + l2$ is the strip length between stands 1 and 2 as in (2.43) and (2.44), and other variables as noted previously. Thus the tension x_1 (2.26) for strip between stands 1 and 2 can be expressed in state-space form as

$$\dot{x}_1 = A_{1,14}x_{14} + A_{1,15}x_{15}. \tag{2.61}$$

Likewise other non-zero elements of the A matrix are the element pairs ($A_{2,15}$ and $A_{2,16}$) for stands 2 and 3, and similarly for equivalent element pairs for the remainder of the mill up to ($A_{6,19}$ and $A_{6,20}$) for stands 6 and 7, with equivalent corresponding variables for these stands.

Elements on the diagonal $A_{7,7}$ through $A_{13,13}$ are each $-\ 1/\tau_S$, and similarly for elements $A_{14,14}$ through $A_{20,20}$ that are each $-\ 1/\tau_V$, and elements $A_{21,21}$ through $A_{26,26}$ that are each $-\ 1/\tau_M$, for the first-order lags in (2.19), (2.20), and (2.23).

Similarly element pairs ($A_{27,14}$ and $A_{27,15}$) to ($A_{32,19}$ and $A_{32,20}$), and ($A_{33,14}$ and $A_{33,15}$) to ($A_{38,19}$ and $A_{38,20}$), are used in relationships respectively for looper positions θ and angular velocities ω, so that all the states in Table 2.1 are considered.

Elements of B:

B is a zero matrix except for elements $B_{7,1}$, $B_{8,2}$, … $B_{13,7}$ that are each $1/\tau_S$, elements $B_{14,8}$, $B_{15,9}$, … $B_{20,14}$ that are each $1/\tau_V$, and $B_{21,15}$, $B_{22,16}$, … $B_{26,20}$ that are each $1/\tau_M$.

Elements of $C(x)$:

$C(x)$ is a zero matrix except for elements that are of the general form $C_{i,j} = p_i/x_j$ where p_i is the variable assigned to y_i (Table 2.1) and x_j is selected to give a reasonable value to the element $C_{i,j}$.

Representation of Mill Characteristics by Models and Plots

The small step changes noted in Sect. 2.4.4 are based on the models presented previously in Sect. 2.1. What follows gives a simple example of how the plots using these perturbations can readily represent the mill as described in the model. This can provide some additional insight into the responses to perturbations in various portions of the mill in an open loop condition as noted in Sect. 2.4.4. This can assist in understanding the dynamics of the process which helps in the development and design of the mill control.

As an example, the reduction of excursions in interstand tensions is important to realize a quality product. In the case of Fig. 2.3, a perturbation in the peripheral speed of the work rolls of stand 4 has resulted in perturbations in the interstand tensions throughout the mill, with the interstand tensions being described by (2.26) and associated relationships. It is of interest to note in the plots that a step change of about + 0.1% in the peripheral speed reference of stand 4 results in about a 17% increase in the peak of the interstand tension between stands 3, 4, and about an equivalent decrease in the interstand tension between stands 4, 5. Thus there is an increase of about 170/1 in the peak response in interstand tension between stands 3, 4 and an equivalent decrease in interstand tension between stands 4, 5.

Similarly the effect on interstand tensions is provided in plots of Fig. 2.5 wherein a step in the torque reference of the looper of stands 3, 4 is considered. In this case a step change of + 1% in the looper torque reference results in about a peak of about + 2% in the interstand tension between stands 3, 4, and about a peak increase of about + 0.3% in the interstand tension between stands 4, 5.

These plots thus readily show the significant difference in the sensitivities in the responses of interstand tensions to a perturbation in stand 4 roll speed and a perturbation in looper torque of stands 3, 4. Thus this is an example of how the plots can provide a useful means of insight into the characteristics of the process dynamics.

References

1. Pittner J, Simaan MA (2010) A useful control model for tandem hot metal strip rolling. IEEE Trans Ind Appl 46(6):2251–2258
2. Pittner J, Simaan MA (2010) An initial model for control of a tandem hot metal strip rolling process. IEEE Trans Ind Appl 46(1):46–53
3. Sims RB (1954) The calculation of roll force and torque in hot rolling mills. In: Proceedings of institution of mechanical engineers, vol 168. UK, pp 191–200
4. Pietrzyk M, Lenard JG (1991) Thermal-mechanical modeling of the flat rolling process. Springer-Verlag, Berlin, pp 12–13
5. Lenard JG, Pietrzyk M, Cser L (1999) Mathematical and physical simulation of hot rolled products. Elsevier, Amsterdam, p 91
6. Ford H, Ellis EF, Bland DR (1951) Cold rolling with strip tension, part I—a new approximate method of calculation and comparison with other methods. J Iron Steel Inst 168:57–72
7. Roberts WL (1983) Hot rolling of steel. Marcel Dekker, New York, p 782

8. Pietrzyk M, Lenard JG (1991) Thermal-mechanical modeling of the flat rolling process, chap 4. Springer-Verlag, Berlin
9. Roberts WL (1983) Hot rolling of steel, Table 22-1. Marcel Dekker, New York
10. Johnson MA, Hearns G, Lee T (2000) The hot strip rolling mill looper: a control study. In: Leondes CT (ed) Mechatronic systems techniques and applications, industrial manufacturing (1). Gordon and Breach, Amsterdam, pp 169–253

Chapter 3
Control Overview

3.1 Background

The tandem rolling of hot metal strip[1] is a highly complex nonlinear process whose characteristics can change considerably during normal and abnormal operations. Over the years various methods of improvement in the control of the mill were developed with various outcomes. This has generally resulted from the upgrading of the structure of the mill, the development of improvements in conventional methods of control,[2] and the development of useful advanced techniques.

During this upgrading of the mill, motorized screwdowns were replaced by hydraulic cylinders, the looper arm driven by an electric motor was replaced by an arm driven by a hydraulic cylinder, plus the addition of reliable sensors of strip tension. Thus, as the mill structure improved and more useful advanced methods of control were adopted, an improvement in the control of the strip tension and thickness began to be realized, with a corresponding improvement in the overall product quality.

There are stringent requirements on the final product which is flat sheets of hot metal that must closely conform throughout to a desired geometry. These requirements dictate that a closed loop control system is needed since, as might be expected, open loop control of the mill is not possible. This is mostly due to undesirable disturbances and uncertainties that must be considered, which often are unpredictable or not easily modeled or measured. Additionally, the control should acceptably handle the following during normal and abnormal conditions, and during threading to assure a quality product over the entire strip length:

---

[1] Herein, and as is often in general usage, the word "strip" will refer to the metal as it is being processed in the mill, and also refer to the mill output.

[2] Conventional control can be understood as a technique that is used in several different mills. Additional detail on what is understood herein as conventional control is provided in Chap. 4.

J. Pittner and M. A. Simaan, *Tandem Hot Metal Rolling Mill Control*, Advances in Industrial Control, https://doi.org/10.1007/978-3-031-58082-6_3

- a wide variety of products whose characteristics can change considerably between various products, and during the processing of a single product;
- stringent customer requirements on the final product;
- a harsh mill environment that restricts the use of certain instrumentation;
- nonlinear strip and mill characteristics that change significantly with temperature and equipment wear, and during normal and abnormal operation;
- long time delays which vary as an element of strip moves between several mill stands;
- tolerance of certain failures of equipment;
- need to promote physical intuition in the design process;
- need to present a user-friendly environment to commissioning, operating, and maintenance personnel, most of whom are less familiar with advanced control techniques;
- reduction of costs by facilitating a rapid start-up;
- others, as might be peculiar to the particular application.

The development of a suitable controller that conforms to the above can be a significant challenge to the control engineer. To help in this effort, several methods are presented herein and in the following chapters that have been shown to conform to these requirements to a reasonable degree. Of course, not every method can be considered as there is an extremely wide array of papers and other publications related to tandem hot mill control. The ones considered herein are deemed typical of various methods and can be considered as a basis for the evaluation of other approaches. It is felt that what is presented will provide a basis for a better understanding of the challenges involved and help promote the development of further advanced methods for improvement.

3.2 Overview of Mill Equipment

In general there are two main areas of interest in the control of the tandem hot metal rolling mill. These are (1) strip profile and flatness in the longitudinal direction, and (2) strip profile and flatness in the lateral direction, over the entire strip length. The main equipment of the mill involved for control in these areas are as follows.

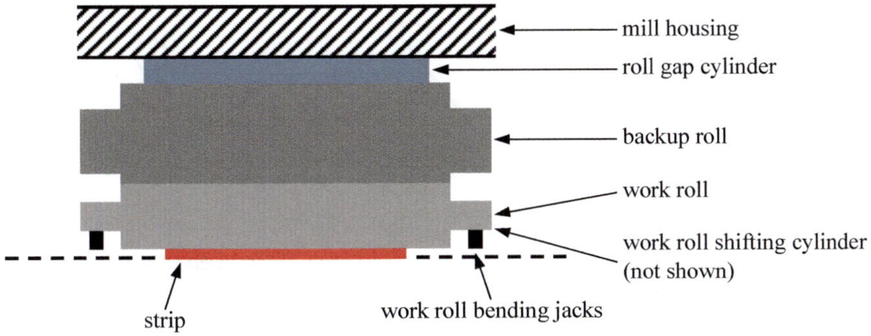

Fig. 3.1 Functional diagram of the upper mill stand area, as when facing the exit of a mill stand[4]

3.2.1 Roll Gap Cylinder

The quality of the product depends very heavily on the control of the hydraulic cylinder (Fig. 3.1) that adjusts the roll gap. The control of this cylinder sets the basic thickness of the strip and also helps to mitigate the effects of various disturbances and uncertainties that could affect the strip thickness and thus ultimately affect the product quality.[3]

3.2.2 Work Roll Bending Jacks

There are usually two work roll bending jacks (Fig. 3.1), one on the drive side and one on the operator side. The work roll shifting cylinders,[5] affect the lateral profile of the strip, which likewise is highly significant for a quality product.

3.2.3 Loopers

The maintaining of nearly constant mass flow of strip between adjacent mill stands and the reduction of excursions in strip tension is crucial to the production of a quality product. The control of the loopers plays an important part in this. The

[3] Some mills have two cylinders, one on the drive side and one on the operator side. A major intent is to provide additional flexibility in the lateral direction.

[5] The hydraulic cylinders for work roll shifting laterally are used to obtain a desired lateral profile (or "crown") of the strip. The initial setting of these cylinders for a particular strip usually remains fixed during rolling of the strip. The use of these cylinders is often denoted as continuously variable crown or CVC.

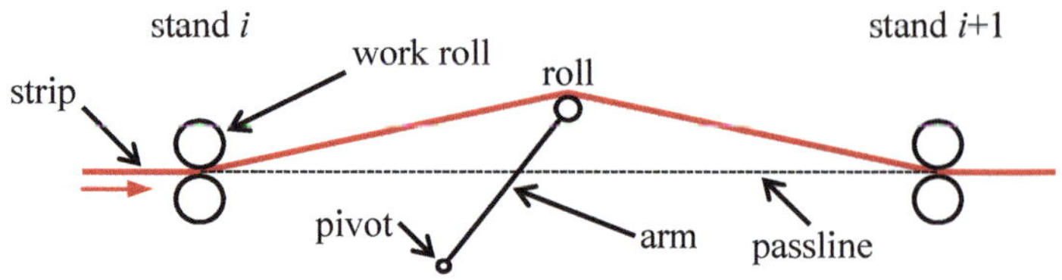

Fig. 3.2 Functional diagram of the looper

general function of a looper (Fig. 3.2) is twofold, i.e., (1) to keep the strip at a desired height between an upstream stand and the adjacent downstream stand; and (2) to maintain constant tension in the strip, ideally free of excursions.

3.3 Overview of Control Challenges

As might be expected, the control of this complex system can present major challenges. As an introduction, some of the challenges to implement the aforementioned requirements are briefly presented in the following overview. More detailed descriptions and how the various challenges are handled by several control concepts are described more fully in the following chapters.

3.3.1 Disturbances

Disturbances are of two categories, i.e., external disturbances and internal disturbances. External disturbances would include, for example, those at the entrance of the first mill stand, such as those in strip thickness, resistance to deformation, and temperature. Internal disturbances would include, but not limited to, roll eccentricities, changes in the material of the strip during rolling, and certain failures of equipment.

External disturbances can result from a slab during cooling at its supports in the reheating furnace.[6] In some instances, additional cooling might result from the slab being near a furnace wall, or other equivalent, so that also there could be a significant temperature gradient across the strip width. After its discharge from the furnace, the slab is processed to be acceptable for its entry into the hot mill. It is generally known that the cooler areas in the slab are usually retained to some degree during its previous processing just before entering the hot mill.

Major internal disturbances include eccentricities in the mill rolls. These are the result of irregularities in the rolls and bearings that cause axial deviations between

[6] The case of disturbances resulting from slabs in a reheating furnace is considered, as it is taken to be more severe than disturbances in the strip characteristics resulting from continuous casting.

the roll neck and the roll barrel, which ultimately result in cyclic deviations in the strip thickness. The backup rolls are a major contributor to this effect, with the work rolls being a lesser contributor.

A lesser effect but with the potential for a cobble (i.e., a wreck in the mill) is a change in the phase[7] of the strip material while it is being processed. Such a change can result from a decrease in the strip temperature during the processing. During the change in the phase, the strip resistance to deformation decreases as the temperature decreases, rather than increasing during the change as might be expected, which must be successfully handled by the control.

Internal disturbances can also result from failures in equipment. The severity of a failure and its effect on the overall process are evaluated on an individual basis as described further in Chap. 5.

3.3.2 Uncertainties

Uncertainties generally are of two categories, i.e., uncertainties in measurement and uncertainties in modeling; and in many instances, an overall deviation from a desired value is a combination of these. An example is the estimate of strip thickness at the exit of a mill stand, especially during threading. Also, there are the uncertainties in modeling wherein linearized models are used to describe certain nonlinear characteristics of the process, as in the following.

3.3.3 Model Mismatch

The tandem hot rolling process is highly nonlinear. In many instances, for simplicity, a general conventional approach is to linearize at a particular operating point and adjust the linearized model as the particular operating point changes. While this can be a simplification in a limited neighborhood of an operating point, additional complexity is often required for significant changes in the operating point during various normal and abnormal operations. This could also increase the magnitude of model uncertainties. More detail is provided in Sect. 3.4.3.

[7] A "phase" of the strip material refers to its metallurgical characteristic; e.g., a change in phase could refer to a change from austenite to ferrite. During the change the material is a mix of austenite and ferrite.

3.3.4 Threading

It is necessary that requirements for a quality product are realized over the entire strip, which includes threading. Additional details are presented in Chaps. 4 and 5.

3.3.5 Tension Excursions

It is generally recognized in the industry that maintaining a nearly constant tension in the strip is necessary for a quality product. This can be a challenge; for example, during threading, wherein the looper is raised from passline to its operating position, which often results in undesirable tension perturbations.

3.3.6 Strip Lateral Characteristics

Much of the conventional control is based on the longitudinal characteristics of the strip at its center line. However, control of the lateral characteristics of the strip is also quite important for a quality product. More detail is available in Chaps. 4 and 5.

3.3.7 Fault Tolerance

A degree of robustness to faults is important to avoid degradation of availability. Typical of these faults are those in the tension or roll force measurements which would preclude a valid measurement of these variables. Such faults could degrade availability and seriously reduce the quality of the output, which could result in a major loss of production and possible equipment damage. Responses to such faults, and others, should be considered.

3.3.8 User Interfaces

It is quite important that interfaces with commissioning, operating, and maintenance personnel are user-friendly and do not require extensive backgrounds in control theory. As noted previously, this is a crucial requirement; especially for advanced control techniques, as it affects costs which strongly depend on a rapid start-up and a continuity of quality production.

3.4 Basic Concepts

Certain basic concepts are generally common to many conventional and certain advanced techniques. These are:

- the "gaugemeter" method for estimating the thickness of the strip as it exits a mill stand;
- consideration of mill roll eccentricities;
- linearization of models;
- feedforward techniques;
- process measurements.

The following provides a brief description of each of these. Some additional others are considered in the chapters herein of the various detailed control strategies.

3.4.1 Gaugemeter (BISRA) Method

This method was developed in the early 1950s by the British Iron and Steel Research Association, thus the notations "BISRA," "gaugemeter," or "gaugemeter (BISRA)" [2]. This method provides a means to estimate the thickness of a strip being processed in a mill stand without requiring a direct measurement. Thus, a reasonably accurate and reliable estimate of the strip thickness is realized without the need for additional instrumentation, which would be more expensive and unable to withstand the harsh mill environment.

The method used in the gaugemeter technique is the measuring of the total roll force and the mill stretch characteristic as depicted in Fig. 3.3. In this characteristic the mill "stretches" nearly linearly, similar to a spring, in accordance with the total load force during rolling. Since most of the rolling occurs in the nearly linear portion of the characteristic, the relationship

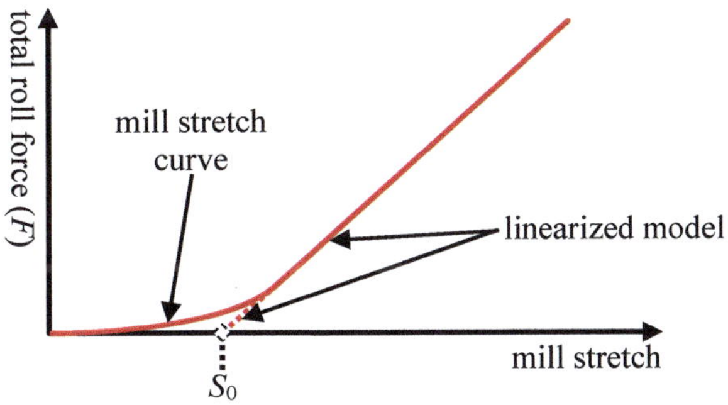

Fig. 3.3 Mill stretch characteristic

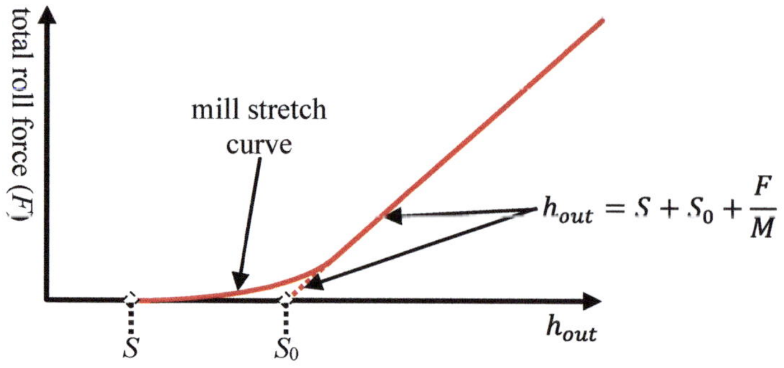

Fig. 3.4 Mill output characteristic

$$\text{mill stretch} = S_0 + \frac{F}{M}, \tag{3.1}$$

is a linear approximation, where S_0 is the extrapolated intercept of the linearized model on the mill stretch axis, F is the total roll force, and M is the mill modulus.

In the case wherein the mill gap is initially at the dimension S, the output thickness h_{out} can be approximated using (3.1) as

$$h_{\text{out}} = S + S_0 + \frac{F}{M}, \tag{3.2}$$

and as depicted in Fig. 3.4.

Usually the mill stretch curve is determined by a calibration following every roll change, wherein the characteristic is approximated in the nonlinear region by a polynomial, and by (3.2) in the linearized area. However, in (3.2) it should be noted that the mill modulus M can have an uncertainty of as much as $\pm$ 20% which should be considered in the control design. This is mostly due to changes in the diameter of the backup rolls caused by mechanical wear and temperature changes during normal operations.

Additionally, it can be noted that a change in the incoming thickness or in the resistance to deformation of the strip, can cause a change in the output thickness. To get a feel for how this might be handled in a very basic manner is depicted in Fig. 3.5.

In the example depicted in Fig. 3.5, the mill is assumed to be open loop; i.e., there is no closed loop control of the roll gap position cylinder; the unloaded roll gap is S_{a}, with an operating characteristic of C_{a}; the operating point is denoted as P_1 with a total roll force of F_1, and an output thickness of $h_{\text{out},1}$. An increase in the resistance to deformation is assumed to occur in the incoming strip, which stretches the mill according to the characteristic C_{a}, so that the total roll force becomes F_2, and the output thickness becomes $h_{\text{out},2}$. For compensation, the roll gap position cylinder is now adjusted to increase the roll force to F_3 so that the output thickness is now returned to $h_{\text{out},1}$, and with an operating characteristic C_{b}, an operating point

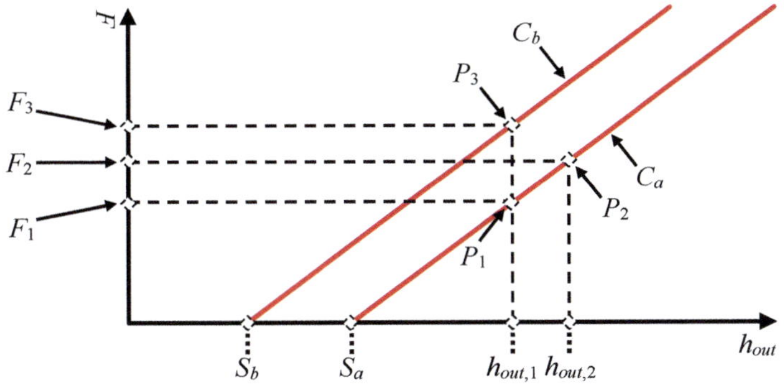

Fig. 3.5 Handling of a change in the incoming strip

of P_3, and an unloaded roll gap of S_b. In this case the mill is considered to be fully compensated, and is sometimes termed to be "infinitely stiff."

3.4.2 *Mill Roll Eccentricities*

Unfortunately, there are eccentricities in the mill rolls which can preclude full compensation. The eccentricity is mostly due to the backup rolls (BURs) since their diameters are greater than the work rolls. In this case for example, an eccentricity in a backup roll can cause a cyclic change in the mill load with a corresponding undesirable change in the strip thickness, which can be significant. This can be further exasperated if there is closed loop control using the BISRA technique as feedback. In this case an increase in the roll force caused by the eccentricity can be seen by the controller as an increase in the strip thickness, when in actuality there is a decrease. Thus, the controller for the strip thickness moves the roll gap position cylinder in the wrong direction which increases the problem.

Moreover, the upper and lower backup rolls usually have slightly different diameters as depicted in mill (b) of Fig. 3.6, and are not always perfectly round as in mill (a) of Fig. 3.6. This difference results in a low-level beat frequency modulation of the eccentricity characteristic, which also must be addressed in the controller. In addition, there are other irregularities in the backup rolls, such as a crude roughness in the surface of the roll that can contribute to the undesirable deviations in the strip thickness. A typical characteristic is depicted in Fig. 3.7.

For these reasons full compensation is usually not implemented, and certain control techniques are used to reduce these effects. Thus, the amount of compensation for eccentricity effects should be carefully considered in a particular application as there could be some imprinting in the strip plus the possibility of excessive wear on the hydraulic system. More detail is provided in Chaps. 4 and 5 for various control methods. Additional detail is also available in [3].

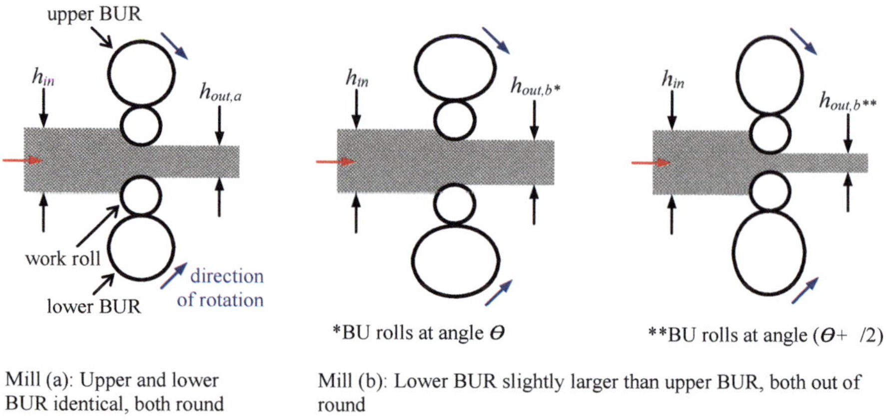

Fig. 3.6 Mill **a** without eccentricity; and mill **b** with eccentricity, where $h_{\text{out,b}}$ changes as BU rolls rotate from angle Θ to $(\Theta + \pi/2)$

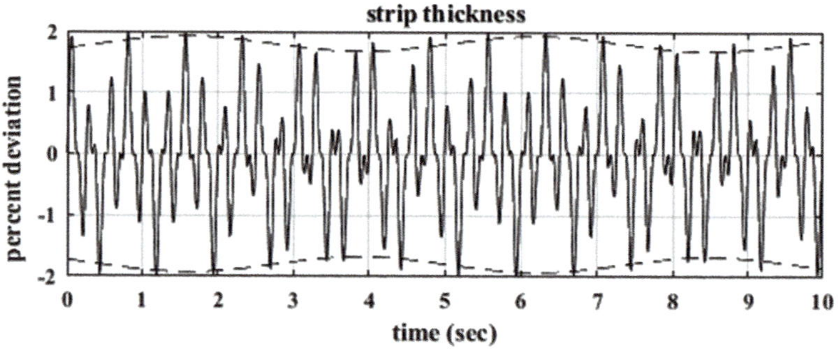

Fig. 3.7 Typical effect of BUR eccentricity on the strip thickness, also depicting a beat frequency effect (dashed lines)[8]

3.4.3 Linearization of Models

In certain methods for the development of a control technique, the linearization of the process model is sometimes used to get an initial feel for the process dynamics, which could be useful in the development of a control method.

In this case the linearization allows the use of a model in linear state-space form, which simplifies the initial insight. The mill is modeled at a single operating point, about which a linear control is developed. The operating point is assumed to be an equilibrium point of the system, so that the only deviations from the point are caused by control action or disturbances.

[8] Adapted from [4]. Copyright 1993. Copyright Marcel-Dekker, courtesy Taylor-Francis (all rights reserved).

Figure developed using MATLAB®.

What follows is an example in the linearization of the mill, wherein the model for the specific roll force is as described in Chap. 2:

$$P = P\big(k, Q_{\mathrm{P}}, \overline{\sigma}, R_{\mathrm{p}}, \delta\big), \tag{3.3}$$

where P (2.1) is the specific roll force, k (2.30) is the resistance to deformation of the material, and considering the temperature of the strip, Q_{p} (2.5) is a factor which compensates for friction and any inhomogeneities of deformation, $\overline{\sigma}$ (2.2) is the mean tension stress of the strip, R_{p} (2.4) is the deformed work roll radius, and δ (2.3) is the stand draft.

In the linearization process, for small excursions around the operating point, the change in the specific roll force is approximated as:

$$\Delta P = \alpha_1 \Delta k + \alpha_2 \Delta Q_{\mathrm{P}} + \alpha_3 \Delta \overline{\sigma} + \alpha_4 \Delta R_{\mathrm{P}} + \alpha_5 \Delta \delta, \tag{3.4}$$

where

$$\alpha_1 = \frac{\partial P}{\partial k};\ \alpha_2 = \frac{\partial P}{\partial Q_{\mathrm{P}}};\ \alpha_3 = \frac{\partial P}{\partial \overline{\sigma}};\ \alpha_4 = \frac{\partial P}{\partial R_{\mathrm{P}}};\ \alpha_5 = \frac{\partial P}{\partial \delta}, \tag{3.5}$$

wherein the partial derivatives are evaluated at the operating point.

In this case, what is shown is intended to be an example of how a linearization might be done at an operating point. However, as noted previously (Sect. 3.3.3), in most instances the actual operation over the entire range is usually significantly nonlinear which must be considered in the controller design.

If a linearized model is to be used, and to avoid excessive model mismatch, significant adjustments to the linearizing coefficients probably are needed to reduce uncertainties as the mill operations change. This could add considerable complexity to the controller. Alternatively, the development of a nonlinear model, or other appropriate method, might be more preferable to reduce complexity and avoid excessive model mismatch. This would have to be evaluated individually for each application. Examples are provided in Chaps. 4 and 5.

3.4.4 Feedforward Techniques

Feedforward techniques have been used (and proposed) to improve responses in various areas of the mill by compensating for disturbances, which is done by tracking a disturbance. The main idea behind these techniques is to design the feedforward strategy such that certain disturbances have only a small effect on the desired output. An example for a linear single-input–single-output (SISO) system is depicted in Fig. 3.8. More detail is available in [5].

In Fig. 3.8, $D(s)$ is the Laplace transform of the disturbance, and is taken to be measurable or well described if unmeasurable; $M(s)$ is the system being controlled;

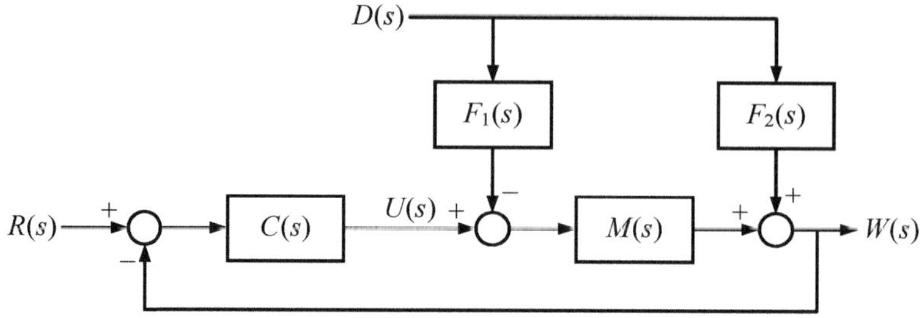

Fig. 3.8 Feedforward control in a linear SISO system for compensation of a disturbance

$C(s)$ is the controller; $W(s)$ is the system output; $R(s)$ is the controller reference; $U(s)$ is the output of the controller; $F_1(s)$ and $F_2(s)$ are to be determined by the feedforward technique to reduce excursions in the system output due to the disturbance. Some basic insight in determining $F_1(s)$ and $F_2(s)$, such that the disturbance has a reduced effect on $W(s)$ is as follows:

Taking $D(s)$ to be zero, the closed loop system output is:

$$W(s) = R(s)\frac{C(s)M(s)}{1 + C(s)M(s)}. \tag{3.6}$$

If it is assumed also that $C(s)$ is designed such that the system response $W(s)$ is acceptable with zero disturbance, then the added excursion $W_d(s)$ on the output $W(s)$ in response to a non-zero disturbance is determined to be:

$$W_d(s) = D(s)\frac{F_2(s) - F_1(s)M(s)}{1 + C(s)M(s)}. \tag{3.7}$$

Thus from (3.7) the effects of the disturbance will be eliminated if:

$$F_2(s) = F_1(s)M(s). \tag{3.8}$$

It should be recognized that the aforementioned feedforward technique must be used with the closed loop feedback control to reduce the effects of any uncertainties in the feedforward signal $D(s)$, or otherwise in the system. The technique as depicted is for a SISO linear system; with appropriate modification it also can be applied to a multi-input–multi-output (MIMO) linear system.[9] Application to nonlinear systems can require more detailed work to be successful, but the overall idea is similar.

Additionally, this technique as shown is intended to reduce deviations in the transient response as well as having zero steady-state error in the output $W(s)$. This is not the case with controllers using compensation based only on the internal model principle, wherein zero steady-state error is realized, but with less control of the transient portion of the response as seen in $W(s)$. This may be acceptable in certain

[9] An example of a MIMO linearized application is given in [7].

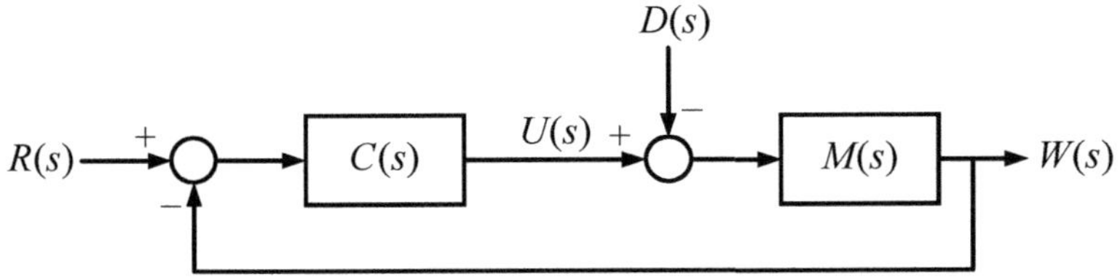

Fig. 3.9 Control in a linear SISO system for compensation of a disturbance using the internal model principle

cases, wherein deviations in the transient response are less crucial, and $D(s)$ can be modified to reduce severe deviations. A diagram of the use of the internal model principle is in Fig. 3.9. More detail regarding the internal model principle is available in [6].

3.4.5 Process Measurements

The harsh environments of the rolling mill areas preclude the desirable measurements of most variables. However, there are some basic measurements that usually are available in most applications and are applicable for control. These include the following as examples:

- position of the hydraulic cylinder;
- hydraulic cylinder force;
- estimates of the speeds of the upper and lower work roll drive shafts;
- angular positions of the looper arms;
- measurements to estimate strip tension at the looper;
- strip thickness at the mill output;
- strip temperature at the mill output;
- others, as applicable for the particular installation.

It can be noted that additional measurements, including some using more advanced technologies, are presently available to attempt to further improve product quality. In general, the usages of these are for specific applications in certain areas, such as measurements of the lateral gradients of temperatures and thicknesses of incoming strip at the first mill stand (e.g., [8]), that must endure the harsh mill environment. The use of more advanced instrumentation in support of certain applications for advanced control is described further in Chap. 5.[10]

[10] Except as otherwise noted, a more detailed treatment of advanced instrumentation is outside the intended scope of this work.

3.5 Additional Concepts

Certain additional concepts are used in more recent conventional methods, and several well-known advanced techniques have been proposed as being useful for control of the tandem hot mill. What follows is a brief description of some of these. Additional details are in Chaps. 4 and 5.

3.5.1 SDRE Technique

The (algebraic) state-dependent Riccati equation (SDRE) technique is not new as it was first suggested in the early 1960s, and developed further for practical applications in later 1990s and earlier 2000s. Eventually later, its user-friendly techniques, with its capability to easily handle nonlinear multi-input–multi-output (MIMO) nonlinear applications, began to be quite popular and useful for many various applications. Some more recent examples are in [9].

In this method an algebraic Riccati equation (ARE) is used to obtain a solution which describes the states that are feedback to the controller of a nonlinear system. The system is described by a state equation whose coefficients can be nonlinear functions of the states. Additionally, the elements of the ARE also can be state-dependent, as needed depending on the model of the system. Thus nonlinearities are easily handled without the need for linearization, which otherwise might be needed to be updated often during actual operation of the process. This, and other beneficial characteristics of this technique, appeared to be quite useful for control of the tandem hot strip mill. The results of recent work as described in Chap. 5 support this.

3.5.2 Model Predictive Control

The function of model predictive control is to determine control inputs by using an optimization criterion that is developed based on a prediction horizon. A system model is developed in state-space form and is usually linear although some nonlinear models have been successfully applied. In most instances a discrete time model is used.

Model predictive control has been used successfully in other applications, particularly in the chemical processing industries, and it appears to offer some advantages, such as possible improvements for nonlinear control. However, it is not clear at present that it has been shown to fully realize a significant improvement beyond some of its presently proposed methods for nonlinear control of the tandem hot mill.

3.5.3 *Sliding-Mode Control*

Sliding-mode control is a method based on a switching function that switches a group of control laws. The switching, as determined by the switching function, is based on the current behavior of the system, so that the system's control law is selected in accordance with some predefined rules that depend on the system's state.

In smaller industrial applications which require robustness to uncertainties in modeling, and with disturbances, sliding-mode control has been quite effective. However, in larger applications for the entire mill it is much less effective. This is because a much more complex application is needed due to many more parameters and undefined uncertainties and disturbances, many of which are also interactive. It is recognized that recently some work has been done that claims improvement. However, to be applicable to the entire area of tandem hot rolling, it appears that additional work needs to be done to be fully useful for larger applications.

3.5.4 H-*Infinity Loop Shaping*

H-infinity (H_∞) control is a method wherein uncertainties and disturbances in linear time invariant multivariable systems are addressed in a systematic manner. In this method, a space (denoted as the Hardy space) consists of the entire group of matrices that represent the stable and proper transfer functions of a particular system. In this system the infinity norm of a particular matrix is minimized, and associated procedures are applied to develop weighting functions; then, using optimization techniques, a stabilizing controller is determined.

The tandem hot mill system is large, highly nonlinear, and with a large range of disturbances and uncertainties that can interact with each other, which makes this method less desirable in most instances. Thus, considerable improvement appears to be required for successful application to the tandem hot mill and would need to be addressed in future investigations.

3.5.5 *Comment*

The aforementioned is intended to give a very brief introduction to some better-known methods. What follows in Chaps. 4 and 5 will present these and some additional methods in more detail.

References

1. Prinz K et al (2018) Optimization-based feedforward control of the strip thickness profile in hot strip rolling. J Process Control 64:100–111
2. British Iron and Steel Research Association. UK patent 713–105
3. Kugi A et al (2000) Active compensation of roll eccentricity in rolling mills. IEEE Trans Ind Appl 36(2):625–632
4. Ginzburg VB (1993) High quality steel rolling, chap 11.1, 11.7. Marcel Dekker, New York
5. Goodwin GC et al (2001) Control system design, chap 10.5. Prentice-Hall, Upper Saddle River
6. Francis BA, Wonham W (1976) The internal model principle of control theory. Automatica 12:457–465
7. Hoshino I et al (1988) Observer-based multi-variable control of the aluminum cold tandem mill. Automatica 24(6):741–754
8. Prinz K et al (2017) Automatic control under laterally asymmetric rolling conditions combined with feedforward. IEEE Trans Ind Appl 53(3):2560–2568
9. Cimen T (2010) Systematic and effective design of nonlinear feedback Riccati equation (SDRE) method. Annu Rev Control 34(1):32–51

Chapter 4
Conventional Control

4.1 Characteristics of Conventional Control

As noted earlier (Chap. 3), the tandem rolling of hot metal strip is a highly complex nonlinear process whose characteristics can change considerably during normal and abnormal operations. Additionally, there are stringent requirements on the final product which is flat sheets of hot metal that must closely conform throughout to a desired geometry, all of which presents a significant challenge to the control engineer. However, if everything was ideal, there would be no need for closed loop control of the mill. Unfortunately, this is not the case, which is why closed loop control, conventional or advanced, is needed.

What is considered herein as conventional control is a common technique that is used in several different mills; or if a certain method that has been generally recognized over several years, it may be understood as conventional control. Also, it can be noted that earlier applications of previous advancements might be considered as conventional. Thus, when a particular method of control is compared with conventional control, to have significant meaning the particular conventional control should be clearly identified. For example, almost all the more recently used (or proposed) methods will generally exhibit major improvement when compared with the earlier conventional control, but may be somewhat less improved when compared with more recent conventional methods. Thus when using this comparison herein, the specific conventional methods are noted.

4.2 Loopers

The looper plays a very important part in the control of the mill, and thus it is considered that an initial insight into its function is helpful for a further understanding of the remainder of the mill control. What follows is intended to present an initial

J. Pittner and M. A. Simaan, *Tandem Hot Metal Rolling Mill Control*, Advances in Industrial Control, https://doi.org/10.1007/978-3-031-58082-6_4

understanding of the control of the looper, and realize a better insight into the overall functions of the rest of the control, which have strong interactions with the loopers.

The major function of the looper is to maintain mass flow between adjacent stands, and significantly reduce excursions in interstand tension. This is highly important for the overall quality in the mill operations and ultimately in the final product of the mill. For example, excessive strip tension can cause narrowing of the width, a reduction in the thickness, wavy edges, and other distortions in the profile, all of which are undesirable. Additionally, it can be noted from the looper model[1] that there is the potential for significant interaction between the looper angle and strip tension so that stability and desired performance might be difficult to realize [1]. This would be especially important in this case wherein even small excursions in the strip tension are undesirable, and similarly wherein excursions in the looper angle are to be avoided.

Some specific requirements are as follows:

- the system should be stable over the entire range of operations of the mill, including normal and abnormal situations;
- undesirable excursions in strip tension and looper angle should be excluded;
- reasonably acceptable performance should be retained in the presence of disturbances and uncertainties;
- others, such as needed to retain the above in both lateral and longitudinal directions.

4.2.1 *Earlier Conventional Control*

Originally the looper was intended for use by the operator in event there was a mismatch of the mass flow between adjacent mill stands. In these uses, the looper was generally kept at or near passline position, with the operator raising the looper only if undesirable conditions occurred. Generally, the result was higher tensions in the strip, with ultimate adverse effects on the geometry of the final product.

Later, the looper was controlled to be raised to an operating position. In the late 1960s and early 1970s control of the looper was improved with the advent of computerized automatic control of rolling mill processes. In that period what became a method of conventional control was the use of an electric torque motor to control the strip tension [2]. The position of the looper was determined by the speed of the strip as set by the speed of the mill rolls, with a control function, as depicted in Fig. 4.1.

---

[1] The applicable areas of the model are given in Appendix Section "Interaction Between Strip Tension Stress and Looper Angle".

As shown in Fig. 4.1, under normal operations the looper position was controlled by the PI to maintain looper position (θ_r), with adjustments as needed by the torque motor to establish desired strip tension (σ_r). In the event of a perturbation in the speed exiting the upstream stand *i*, the change in speed changes the looper position, which is measured by the angle (θ). The change in (θ) is fed back to the PI, which provides a speed trim signal (δ) to the motor drives of rolls of stand *i*. The trim signal adjusts the speed of the strip exiting the stand, which adjusts the position of the looper and its angle (θ). The change in the angle is also fed to the torque calculator, which computes the estimated required torque to the torque motor drive, to adjust the strip tension to its desired value, which is estimated based on the angle (θ), the referenced desired tension (σ_r), strip thickness (h_r), and width (b_r).

Similarly, if there is a small perturbation in tension, the looper angle (θ) is changed due to the unbalance between looper torque and tension. The PI then changes the speed trim to maintain (θ_r), and the torque calculation adjusts the looper torque accordingly.

In an initial evaluation, what is depicted in Fig. 4.1 may have appeared acceptable. However, further evaluation, simulation, and actual usage would reveal some characteristics that are less desirable. A main issue was the interaction between

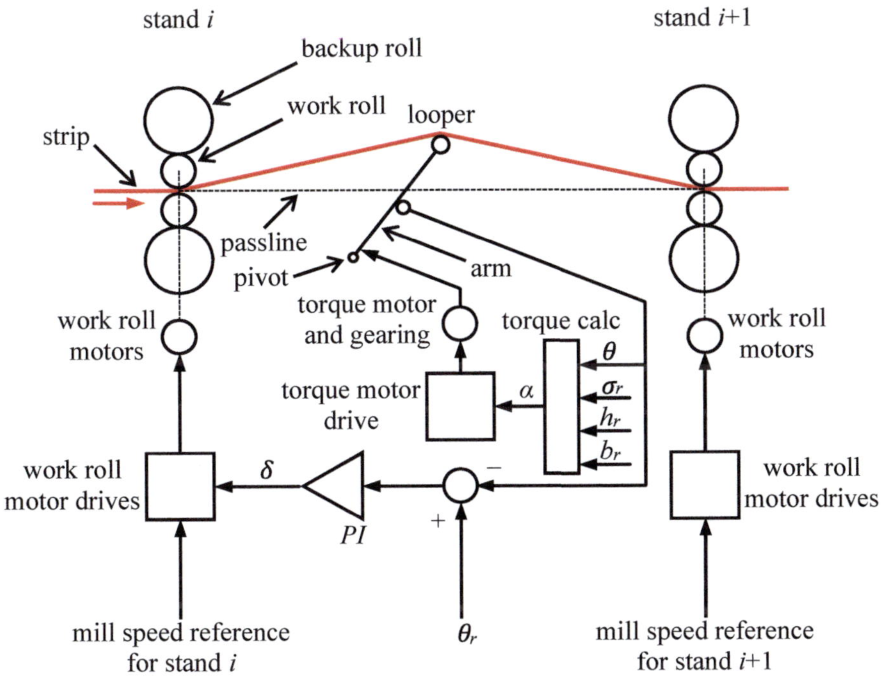

Fig. 4.1 Functional diagram of the looper with earlier conventional controller. α: torque signal; δ: speed trim signal; PI: proportional–integral control; θ: looper angle (measured); θ_r: looper angle (reference); σ_r: strip tension (reference); h_r: strip thickness (reference); b_r: strip width (reference)

the characteristics of the control loop for looper position and those of the torque controller. With this arrangement, both of these functions have slow responses. The relationship between interstand strip tension stress and the looper angle is further described in Appendix Section "Interaction Between Strip Tension Stress and Looper Angle".

For example, the response of the control loop that uses a trim on the speed of the work roll drives is a somewhat slow correction to a deviation in the looper angle. In the case of the torque, the inertia of the torque motor and gearing precludes a more rapid change in the looper torque, and thus in the strip tension. Thus, the responses of the looper angle and the strip tension are both slow and interacting, and possibly unstable if any of the gains of the control loops are increased. A faster response is especially desired in the tension, which would be expected to be held nearly constant during changes in the looper angle.

4.2.2 Improvement in Conventional Control

What was described previously was later updated to replace the torque motor and gearing with a hydraulic cylinder, plus the addition of a measurement to better estimate the strip tension. Figure 4.2 describes this change.

As in the case shown in Fig. 4.1, the looper position is controlled by the PI to maintain the looper position (θ_r), with the speed of the upstream mill stand being trimmed to adjust the looper angle in response to a disturbance. Thus the speed of the response of the position loop to a disturbance does not change considerably from that of Fig. 4.1.

However, in the case of Fig. 4.2, the response of the closed loop control of the strip tension is faster than the response of the looper position loop, so that excursions in the tension are considerably reduced during the response to a disturbance. This reduced the interaction of the loops and with a faster response in the tension. This resulted mostly from the replacement of the torque motor and gearing with the hydraulic cylinder which reduced the inertia in the tension loop; and also, the addition of a measurement improved the estimate of the tension.

The overall result was considered beneficial as it reduced excursions in the strip tension, which was deemed important in producing a quality product, while some minor excursions in the looper position also remained.

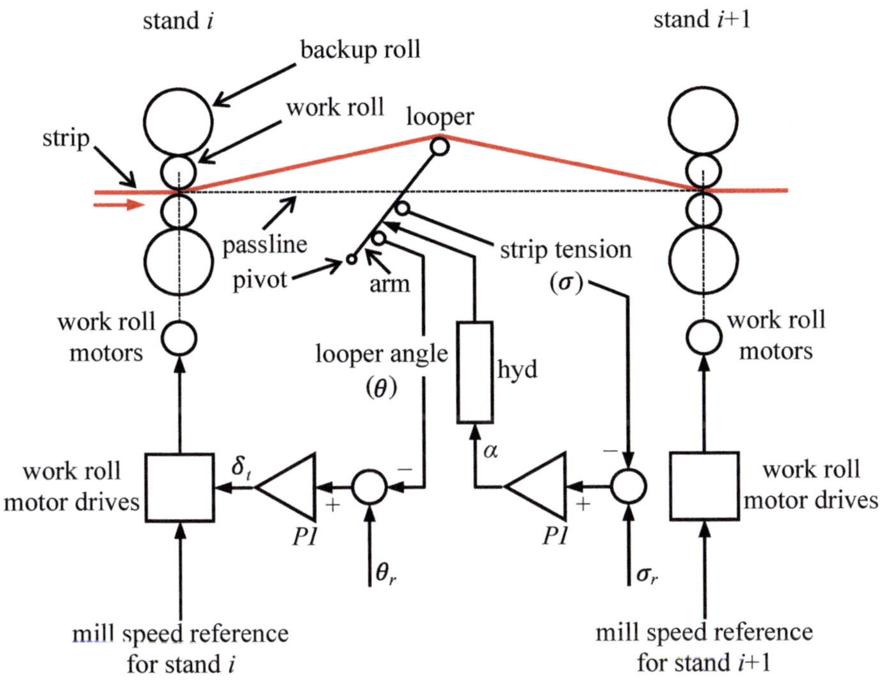

Fig. 4.2 Functional diagram of the looper with modified earlier conventional controller. PI: proportional–integral control; hyd: hydraulic/control; δ_t: speed trim signal; α: tension control signal; θ_r: looper angle (reference); θ: looper angle (measured); σ_r: strip tension (reference); σ: strip tension (measured)

4.2.3 *Further Improvement*

In the following years many further improvements in conventional control of loopers were proposed and in many instances, successfully implemented to a reasonable degree. Some typical examples are as follows.

4.2.3.1 Further Improvement, Example 1

In this example, in the early 1980s a multivariable non-interacting control (NRC) technique began to be applied. The intent was to significantly reduce the interactions between the strip tension and looper position. This technique was based on [3], and was applied for example in [4, 5],[2] to reasonably reduce interference between these two functions.

In these applications a linear control method is based on a linearized model. The main concept of NRC is to add a compensating function to reduce the cross coupling

[2] Figure 4.3 is a typical functional example.

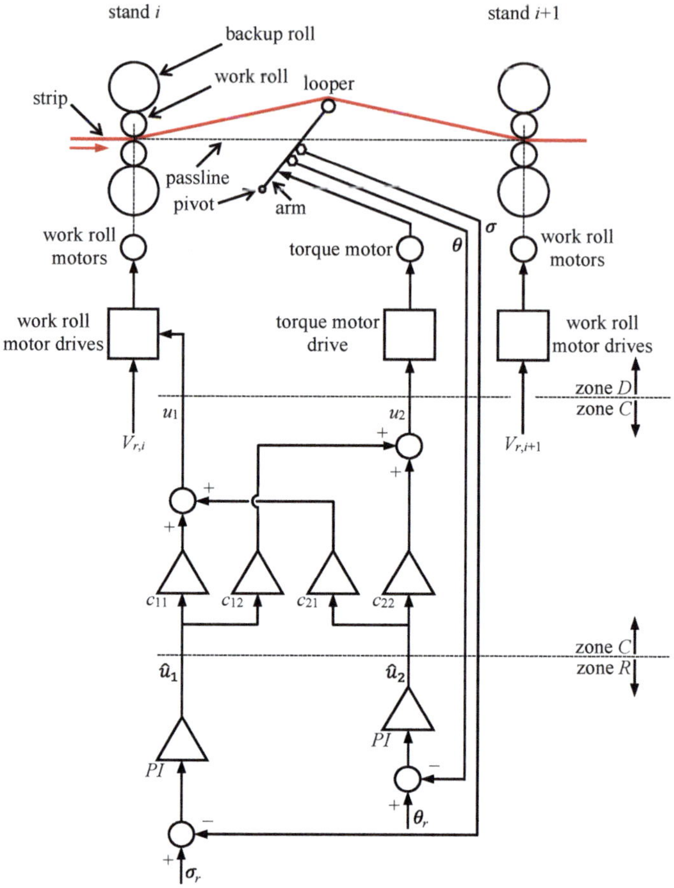

Fig. 4.3 Functional diagram of the conventional looper controller using non-interactive multivariable control and direct drive torque motor with reduced inertia, including measurements of looper angle and strip tension. PI: proportional–integral, main control; c_{11}, c_{12}, c_{21}, c_{22}: transfer functions, cross control; $V_{r,i}$: mill speed ref, stand i; $V_{r,i+1}$: mill speed ref, stand $i+1$; σ_r: strip tension (reference); σ: strip tension (measured); θ_r: looper angle (reference); θ: looper angle (measured); $\hat{u}_1$: output of main control, strip tension; $\hat{u}_2$: output of main control, looper angle; u_1: output of cross control, mill speed trim*; u_2: output of cross control, torque motor drive ref; * work roll drives invert u_1 such that an increase in u_1 causes an increase in σ

between strip tension and looper position. This was done by adding a matrix to modify the controller such that the control action for the tension affects only the tension, and similarly for the looper position. For example, a change in the desired tension (σ_r) would correspondingly change (σ) with only a negligible deviation in looper angle (θ).

The addition of a compensating matrix would result in an overall diagonal matrix $Q(s)$, such as in (4.1). In this case each of the outputs (σ) and (θ) depends only on a single matrix element, $q_{11}(s)$ or $q_{22}(s)$, where $\hat{u}_1$ and $\hat{u}_2$ are the inputs for (σ) and

(θ), and s is the Laplace transform:

$$\begin{bmatrix} \sigma \\ \theta \end{bmatrix} = \begin{bmatrix} q_{11}(s) & 0 \\ 0 & q_{22}(s) \end{bmatrix} \begin{bmatrix} \hat{u}_1 \\ \hat{u}_2 \end{bmatrix}. \tag{4.1}$$

This technique is applied functionally as in Fig. 4.3 wherein there are three zones. Zone D includes the equipment being controlled by variables u_1 and u_2 that modify the strip tension and the looper position. Zone C includes transfer functions that are elements c_{11}, c_{12}, c_{21}, and c_{22} of the compensating matrix C. This matrix modifies the variables u_1 and u_2 such that only the desired tension (σ) or desired looper angle (θ) is affected. Zone R includes the PI main controllers for (σ_r) and (θ_r) that provide input signals $\hat{u}_1$ and $\hat{u}_2$ to matrix C. The development of the matrices for zones D and C, and the overall results, are as follows.

In the case of zone D the linearized model of the equipment under consideration is somewhat complicated. Thus a simplified model[3] of zone D in (4.2) is assumed to help give initial insight into some of the basic ideas of the NRC method:

$$\begin{bmatrix} \sigma \\ \theta \end{bmatrix} = \begin{bmatrix} d_{11}(s) & d_{12}(s) \\ d_{21}(s) & d_{22}(s) \end{bmatrix} \begin{bmatrix} u_1 \\ u_2 \end{bmatrix}. \tag{4.2}$$

Similarly in zone C, and including zone C with $\hat{u}_1$ and $\hat{u}_2$:

$$\begin{bmatrix} \sigma \\ \theta \end{bmatrix} = \begin{bmatrix} d_{11}(s) & d_{12}(s) \\ d_{21}(s) & d_{22}(s) \end{bmatrix} \begin{bmatrix} c_{11}(s) & c_{12}(s) \\ c_{21}(s) & c_{22}(s) \end{bmatrix} \begin{bmatrix} \hat{u}_1 \\ \hat{u}_2 \end{bmatrix}. \tag{4.3}$$

Combining C and D and diagonalizing:

$$\begin{bmatrix} \sigma \\ \theta \end{bmatrix} = \begin{bmatrix} d_{11}(s)c_{11}(s) + d_{12}(s)c_{21}(s) & 0 \\ 0 & d_{21}(s)c_{12}(s) + d_{22}(s)c_{22}(s) \end{bmatrix} \begin{bmatrix} \hat{u}_1 \\ \hat{u}_2 \end{bmatrix}, \tag{4.4}$$

and C can be determined as:

$$C(s) = \begin{bmatrix} c_{11}(s) & \dfrac{-d_{12}(s)c_{22}(s)}{d_{11}(s)} \\ \dfrac{-d_{21}(s)c_{11}(s)}{d_{22}(s)} & c_{22}(s) \end{bmatrix}. \tag{4.5}$$

In certain applications, for simplification the effects of frequency were taken as less significant. Thus, ideally at zero frequency, D becomes a constant matrix, C then becomes a constant matrix (4.6), and (4.4) becomes the identity matrix as in (4.7),

$$C = \begin{bmatrix} \dfrac{d_{22}}{d_{11}d_{22} - d_{12}d_{21}} & \dfrac{d_{12}}{d_{21}d_{12} - d_{11}d_{22}} \\ \dfrac{d_{21}}{d_{12}d_{22} - d_{11}d_{12}} & \dfrac{d_{11}}{d_{11}d_{22} - d_{21}d_{12}} \end{bmatrix}, \tag{4.6}$$

[3] A more detailed model is available in [4, 5].

and

$$\begin{bmatrix} \sigma \\ \theta \end{bmatrix} = \begin{bmatrix} 1 & 0 \\ 0 & 1 \end{bmatrix} \begin{bmatrix} \hat{u}_1 \\ \hat{u}_2 \end{bmatrix}. \tag{4.7}$$

While some specific detail on actual performance is a bit sparse, it is generally recognized that the NRC technique had significant advantages over the earlier conventional methods. A main advantage is the significant improvement in the response to various normal and certain abnormal situations. This has resulted mostly from the addition of measured strip tension as feedback, the reduction of inertia in the drive shaft of the looper, and the significant reduction of interaction between tension and looper angle due to a non-interacting control structure.

However in some instances larger perturbations in tension such as in the processing of thin strips, the response is often slow and unacceptable [6]. This is mostly due to the slower response of the main motor drive speed controller. Additionally, the use of a linear controller can cause certain other degradations in performance due to model mismatch with the nonlinear system. This must be addressed to avoid undesirable qualities in the final product.

4.2.3.2 Further Improvement, Example 2

In this example, which was implemented in the later 1980s and early 1990s, an optimal multivariable control technique with an integrator, and direct drive motor with reduced inertia, was utilized. The looper motor was coupled directly to the looper without gearing to further reduce the undesirable effects of inertia. Additionally, instrumentation to measure strip tension, looper position, and looper motor speed was included. It was considered that the multivariable technique would better handle larger disturbances in the tension, and with the integrator to realize zero steady-state error in strip tension and looper angle. Figure 4.4 presents a typical system based on [6] and the references cited therein.

The following provides some insight into the characteristics of the control, which is based on a nonlinear model that is linearized at an operating point. The system shown in Fig. 4.4 can be described in state-space form as

$$\frac{\mathrm{d}x}{\mathrm{d}t} = Ax + Bu, \quad x(0) = x_0, \tag{4.8}$$

$$y = Cx, \tag{4.9}$$

where

$$x = \begin{bmatrix} \sigma \ \omega \ \theta \end{bmatrix}', \ y = \begin{bmatrix} \sigma \ \theta \end{bmatrix}', \ u = \begin{bmatrix} \delta_t \ \omega_r \end{bmatrix}', \tag{4.10}$$

the symbol ($'$) indicates the matrix transpose, and where

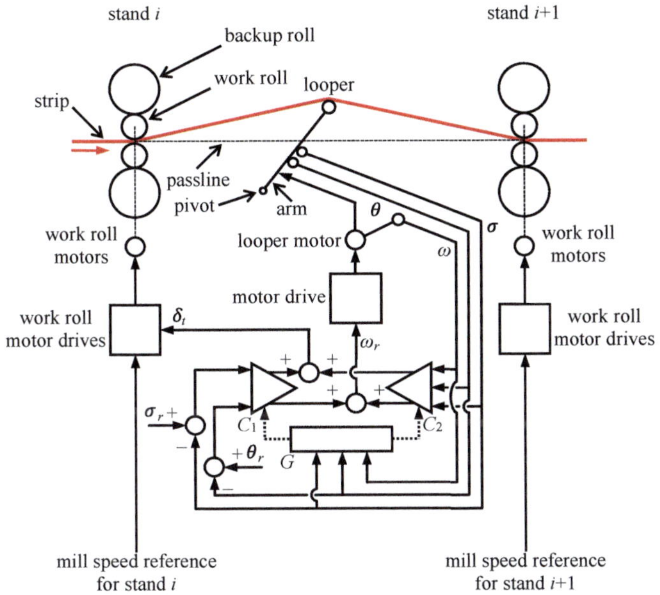

Fig. 4.4 Functional diagram of the looper with conventional controller using multivariable optimal control with an integrator, and direct drive motor with reduced inertia. C_1: integral control; C_2: state-feedback control; G: gain adjust (C_1, C_2); σ_r: strip tension (reference); σ: strip tension (measured); θ_r: looper angle (reference); θ: looper angle (measured); ω_r: looper motor speed (reference); ω: looper motor speed (measured); δ_t: work roll motors speed trim*; * work roll drives invert δ_t such that an increase in δ_t causes an increase in σ

$$A = \begin{bmatrix} a_{11} & a_{12} & 0 \\ 0 & a_{22} & 0 \\ 0 & a_{32} & 0 \end{bmatrix}, \quad B = \begin{bmatrix} b_{11} & 0 \\ 0 & b_{22} \\ 0 & 0 \end{bmatrix}, \quad \text{and} \quad C = \begin{bmatrix} 1 & 0 & 0 \\ 0 & 0 & 1 \end{bmatrix} \tag{4.11}$$

are determined by the linearization, wherein the elements of each matrix are determined based on the system model, similar to [6].

As described in Appendix Section "Derivation of Optimal Control", the state Eq. (4.8) is rearranged (4.38) to realize a utilization of the algebraic Riccati equation (ARE) (4.42) for optimality, and the addition of an integrator. The results (4.49) for general application are:

$$u = -R^{-1}B'K_{21}\int_{t_0}^{t}(y_r - y)\mathrm{d}t - R^{-1}B'K_{22}\int_{x(t_0)}^{x(t)}\mathrm{d}x + U(t_0), \tag{4.12}$$

where R is an element of the ARE performance index (4.39), $K = \begin{bmatrix} K_{11} & K_{12} \\ K_{21} & K_{22} \end{bmatrix}$ is the solution of the ARE, and $U(t_0)$ is the initial condition[4] of the associated differential Eq. (4.48) of Appendix Section "Derivation of Optimal Control". The specific application of (4.12) to the characteristics of the system mechanics using control functions C_1 and C_2 (Fig. 4.4) is:

$$u_{C1} = -R^{-1}B'K_{21}\int_{t_0}^{t}(y_r - y)dt \tag{4.13}$$

$$u_{C2} = -R^{-1}B'K_{22}(x(t) - x(t_0)), \tag{4.14}$$

and

$$u = u_{C1} + u_{C2} + U(t_0). \tag{4.15}$$

The gain adjust function G (Fig. 4.4) allows for a table of adjustments to the ARE to compensate for model mismatch due to nonlinearities of the linearized system. For example, in [6] elements Q and R of the ARE (4.42) are adjusted to compensate somewhat for model mismatch depending on the variations in the nonlinearities and their variations during actual processing.

The system was installed on an actual mill as described in [6]. Some of the major advantages over previous methods (e.g., earlier conventional [2], and non-interactive [5]) were as follows:

- improved response during interaction with higher performing automatic gauge control (AGC) on main mill stands;
- better performance on rolling of thin strips, strips of higher temperature, higher carbon steels, and higher speed rolling;
- better response to disturbances such those caused by skid marks on incoming strips;
- additional others.

An overall improvement in tension and angle deviations is summarized in Table 4.1 (data derived from [6]).

Table 4.1 Standard deviations in excursions from desired strip tensions and looper angles, compared to other controls

Control	Standard deviation	
	Tension	Angle
Earlier conventional	1.0	1.0
Non-interactive	0.83	0.29
Optimal multivariable	0.45	0.24

[4] $U(t_0)$ is implied in Fig. 4.4.

As described in the aforementioned, the optimal multivariable control has produced improvements over previous conventional methods. However, most of these improvements have resulted from the use of a table of the Q and R matrices to account for major mismatches between the linearized model and the actual mill. In this approach, an operation that was outside the useful area of the linearized system required the replacement of these matrices by those in the table to preclude an undesirable response.

While this approach has resulted in improvement in several areas, related information describing how the nonlinear conditions were addressed was scant. For example, specific detail on how the matrices Q and R in the table were determined, how performance using these matrices was validated, and other issues of concern generally occurring in using a linearized system for a highly nonlinear process. It is felt that further work using a nonlinear approach, or other applicable method, might have resulted in a simpler system that offered further improvement over the full range of final product.

4.2.3.3 Further Improvement, Example 3

In this example a controller [7] was implemented based on H-infinity (H_∞) techniques. These techniques became quite popular in the later 1990s and earlier 2010s, and were investigated as possible useful methods for improvement in various applications.

In the case of loopers in hot metal rolling mills, it was considered that the application of H_∞ methods would provide additional improvements beyond those already realized by conventional non-interacting control such as [4, 5]. Among the areas of particular interest were improvements in response to disturbances such as perturbations in thickness of incoming strip caused by skid chill marks from the reheating furnace, interactions with recently improved automatic gauge (strip thickness) controllers at mill stands, plus others.

As an example [7], an H_∞ mixed sensitivity method[5] was considered for improving the performance of the controller over a reasonably wide frequency range. A strong rejection of disturbances at the lower frequencies, with reduced effects of noise and uncertainties at the higher frequencies was expected.

However, this method proved to be less desirable for actual operations in the plant as it was less capable of handling uncertainties, the changes in poles and zeros possibly due to system nonlinearities, mismatches in modeling during processing of a wide variety of products, and others.

---

[5] A brief overview of this method is provided in Appendix Section "Brief Overview of H_∞ Methods".

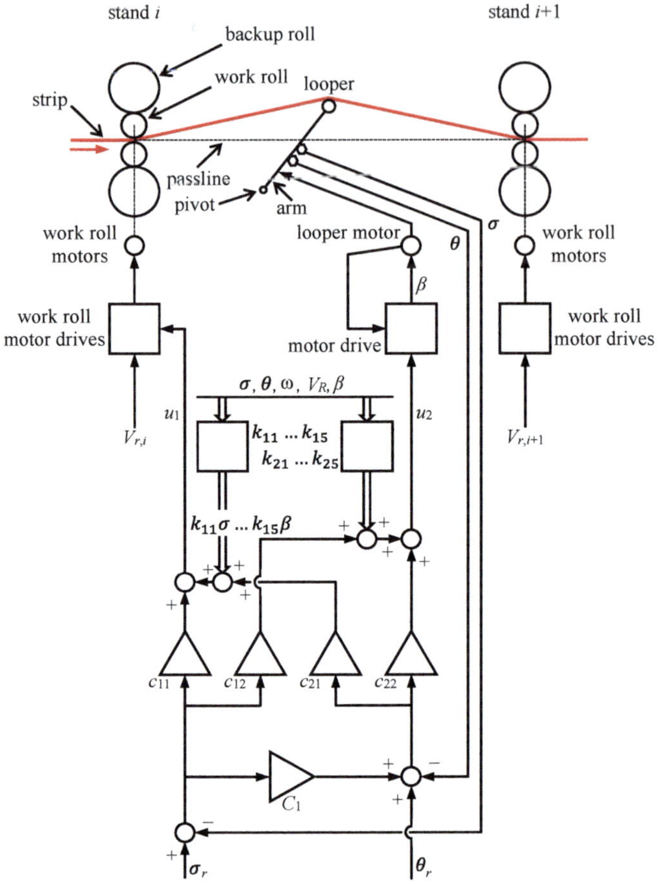

Fig. 4.5 Functional diagram of the conventional looper controller using non-interactive H_∞ multivariable control, including measurements of looper angle and strip tension. σ_r: strip tension (reference); σ: strip tension (measured); θ_r: looper angle (reference); θ: looper angle (measured); ω: looper motor speed; V_R: work roll speed; $V_{r,i}$: mill speed ref, stand i; $V_{r,i+1}$: mill speed ref, stand $i + 1$; u_1: output of control, mill speed trim*; u_2: output of control, looper speed reference; C_1: trim (tension/looper angle); c_{11}, c_{12}, c_{21}, c_{22}: cross control transfer functions (integrators); $k_{11} \ldots k_{15}$, $k_{21} \ldots k_{25}$: constants; β: output of integral function of PI speed controller in looper motor drive; * work roll drives invert u_1 such that an increase in u_1 causes an increase in σ

Additionally, the complexity of the design made on-site tuning difficult. As a revision,[6] a lower order H_∞ state-feedback controller was implemented which did not require cancelation of poles and zeros, and was a bit easier to tune, as depicted in Fig. 4.5, except without constant C_1. This revision was based on the use of state feedback and the solution of an algebraic Riccati equation to determine the gains of the state feedbacks and integrals (i.e., the gains k and c in Fig. 4.5).

[6] Additional detail is available in [7].

Table 4.2 Comparison of tension fluctuations

Control	Tension fluctuation
Non-interactive	1.00
H-infinity	0.67

Unfortunately, when performance was compared to that of conventional non-interacting control, little improvement was realized, especially in the deviations of the strip tension, which was a main objective of the improvement in the controller. An intuitive response was the addition of the constant C_1 in Fig. 4.5, which provided a trim in the looper angle using the error between the measured tension and the tension reference. The reasonably successful ultimate result is depicted in Table 4.2 (data derived from [7]).

The result as depicted in Table 4.2 includes the capability for successfully reducing undesirable deviations in tension from skid marks in the incoming material, which is a significant improvement over NIC methods. This was confirmed for the actual processing of many strips. However, it is not clear how well the performance was improved over a broader range of products and events, and with nonlinear operational conditions, especially since the control was based on a linear concept; and particularly since the major improvement resulted from the intuitive addition of the C_1 constant, rather than the intended strategy of the controller. It is not clear how this constant was determined, or how it might require adjustment during various phases of operation, or if some simplification of the remainder of the control could be realized, or in other improvements.

Moreover, the use of the linear H_∞ technique is fairly complex, and thus its application to the various regimes of the nonlinear system would require additional complexity to verify stability and performance over the entire range of operation. However, even if linearity could be retained, the determination of the weights (Fig. 4.19, Appendix Section "Brief Overview of H_∞ Methods") is not straightforward and would add further complexity, and most likely be more difficult to use by design and operational personnel.

4.3 Strip Thickness Control

The overall control of strip thickness involves responses to a large variety of disturbances, uncertainties, and other possible situations that must be addressed by the control function, such as those described in [8]. To help provide a background for a better understanding of the control challenges involved, what follows describes some of these as examples as considered in certain significant successful earlier conventional control methods. Additional conventional works are described in [9, 10]. More advanced methods are addressed in Chap. 5.

4.3.1 Automatic Gauge Control

A main part of strip thickness conventional control is the automatic gauge control (AGC) function, which has been in use successfully for several years (e.g., [11]). This function is based on the estimate of the output strip thickness of the mill stand using the measurements of the stand roll force and the position of the stand hydraulic cylinder; and also, this estimate is supplemented with measurement of strip thickness using an X-ray monitor at the mill exit or at one located at a midway position in the mill. The very basic functional operation of the AGC is described in what follows.

As depicted in Figs. 4.6 and 4.7 the thickness of a strip exiting the mill stand is h_{out}. As noted in these figures the measured position of the hydraulic cylinder is S_{m}, which applies force F_{roll} to the work rolls via the backup roll to adjust the strip thickness (4.16). The roll force also can be affected by the force F_{wrb} of the work roll jacks (4.17), which are used primarily to adjust the strip lateral characteristics. The mill housing is stretched by the force F_{h} so that the strip output thickness is determined by S_{m} and the roll force[7] F_{roll}/M, where M is the mill modulus, and the estimated output thickness is:

$$h_{\mathrm{out,e}} = \frac{F_{\mathrm{roll}}}{M} - S_{\mathrm{m}}, \tag{4.16}$$

where

$$F_{\mathrm{roll}} = F_{\mathrm{h}} - 2F_{\mathrm{wrb}}. \tag{4.17}$$

Also, it can be noted that during the rolling process the mill stretch can change if the resistance to deformation (hardness) of the incoming strip increases or decreases, which could result in an increased uncertainty in the estimation of the strip output thickness. This additionally would need to be addressed by the gauge control function.

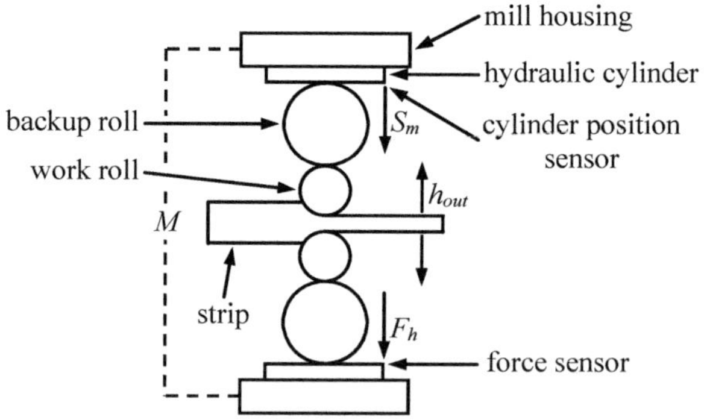

Fig. 4.6 Description of mill stand during rolling

[7] The measurements of F_{h}, F_{roll}, and F_{wrb} are available in most installations.

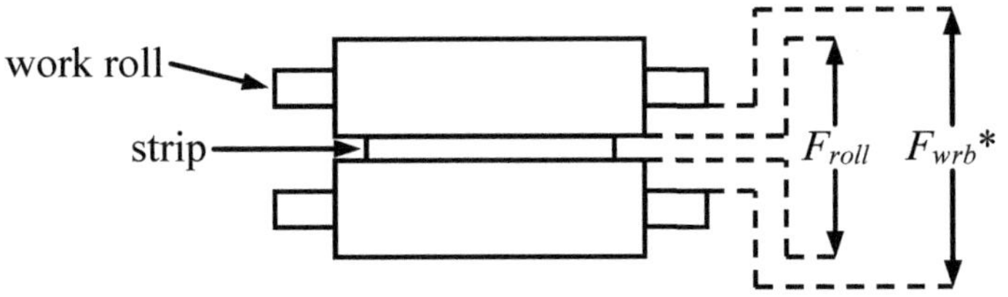

Fig. 4.7 Description of mill work rolls and strip, with work roll bending force F_{wrb} and roll force F_{roll}. * F_{wrb} is by work roll bending jacks (not shown)

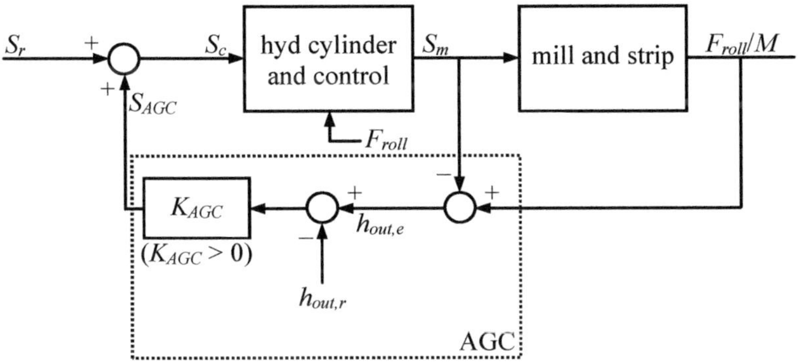

Fig. 4.8 Diagram of the basic AGC

The estimation using this method to predict h_{out} is denoted as the BISRA technique,[8] or also known as the "gaugemeter" method, which is the basis of the (AGC) function.

In general, the basic AGC is depicted as in Fig. 4.8. The work rolls and strip are as in Fig. 4.6. The hydraulic cylinder and control are similar to that described in [12] wherein the output effect on the strip thickness is S_{m}, which is determined by closed loop control of the input reference S_{c}. This control method has been shown in actual use to be quite effective so that essentially $S_{\text{m}} = S_{\text{c}}$.

There are limitations on the gain of K_{AGC}. These limitations are necessary to reduce the likelihood of instability, reduce the effects due to roll eccentricities, or other possible undesirable effects.[9] Also, since the control is proportional, the limitations on the gain will limit the capability for reduction of the error in $h_{\text{out,e}}$.

A simplified example follows to provide some insight into the proportional control of the basic AGC (Fig. 4.8) to handle a change in the incoming strip thickness:

- Initially it is assumed that there is strip in the operating mill, there are no uncertainties or disturbances, the control is open loop, and S_{r} is set such that the estimated

[8] This method to estimate strip thickness was developed in the 1950s by the British Iron and Steel Research Association (BISRA) [13]. More detail is available in Chap. 3.

[9] More detail on the determination of K_{AGC} and its limitations are available in [14].

Fig. 4.9 Simplified functional operation of the basic AGC assuming a linearized mill characteristic

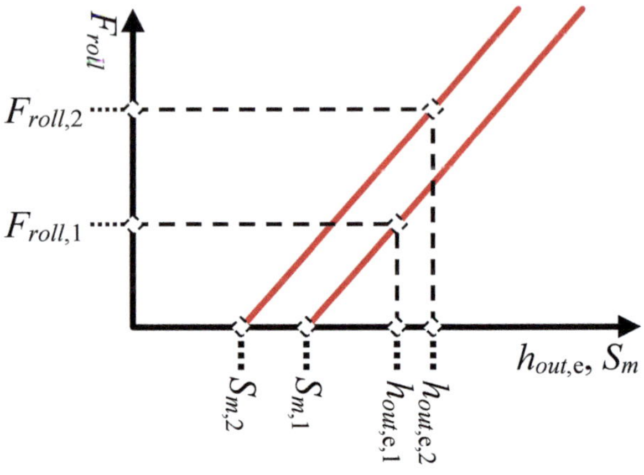

output thickness ($h_{\text{out,e}} = F_{\text{roll}}/M - S_{\text{m}}$) equals the desired thickness $h_{\text{out,r}}$, i.e., S_{AGC} is zero and $S_{\text{m}} = S_{\text{c}} = S_{\text{r}}$. The mill operating point is ($h_{\text{out,e,1}}$, $F_{\text{roll,1}}$), as in Fig. 4.9.

- The control loop then is closed, and since there are no changes in the system, S_{AGC} remains at zero.
- The incoming strip thickness then is assumed to increase so that the roll force F_{roll} increases with a corresponding increase in $h_{\text{out,e}}$.
- The increase in $h_{\text{out,e}}$ causes an increasing in S_{AGC} and thus in S_{c} and S_{m}.
- The increasing in S_{m} then adjusts the change in $h_{\text{out,e}}$ and S_{AGC} to a new operating point, consistent with the change in F_{roll}.
- However the change in $h_{\text{out,e}}$ and S_{AGC} is in accordance with proportional control characteristics, so that a slight error remains in the estimate of $h_{\text{out,e}}$, and the resulting operating point is ($h_{\text{out,e,2}}$, $F_{\text{roll,2}}$).

4.3.1.1 Mill Stand Effects

As might be expected, there is more to the AGC than just the basic controller. Considerable experience has shown that the BISRA method has been working well. However, some issues based on changes in the characteristics of the mill stand itself can increase undesirable changes in the roll force. As an example, in the case of previous earlier work,[10] some of these are the increases in roll temperatures during initial start-up, changes in roll bearing characteristics during changes in the mill speed, effects of work roll bending, and of work roll eccentricity. Some of these typical cases and the responses to them are as follows.

The increase in work roll diameters due to an increase in roll temperatures can result in an increase in roll force. This increase in diameters actually results in a decrease in the actual strip exit thickness as follows. The AGC would then interpret

[10] More detail is provided in [8].

this (as in (4.16) and Fig. 4.9) as an increase in the mill stretch (F_{roll}/M), and thus an increase in the estimated strip thickness ($h_{out,e}$). The AGC would then respond by increasing (S_{AGC}), the cylinder position (S_m), and thus further decreasing the actual strip exit thickness. This was addressed by the development of compensation functions. These were based on measurements of roll temperatures at each mill stand during and after the initial start-up, and were applied with reasonable success.

Similarly, compensation functions were developed to account for roll bearing oil film effects that could affect the roll gap during different mill speeds. This effect occurs when the mill speed is increased such that the roll bearings tend to move more toward the center. This causes the thickness of the exit strip to decrease, and the subsequent roll force to increase, which is misinterpreted by the AGC so that the error is actually increased. This was corrected by operating the mill open loop without strip and taking measurements that could be used in the modifications of the mill stretch curve,[11] which generally is taken after each roll change.

The work roll bending forces are used to modify the lateral shape of the strip and to reduce the strip crown. These forces can also affect the gauge of the strip (as in (4.17)). The effect of these forces depends on the width of the strip and can require possible recalibration depending on the characteristics of the strip, which was reasonably effective in these cases.

Work roll eccentricity can affect the strip thickness as noted in Chap. 3. Control methods for reducing its effects are described and addressed more fully in Chap. 5.

4.3.1.2 Absolute Gaugemeter Control

As noted previously, there is an inherent error in the strip thickness estimate using the proportional method of the AGC control. Additionally, during threading a deviation in the expected strip hardness could likewise affect the strip head end thickness at the output of the roll gap. This would not be fully detected until measurement by an X-ray gauge located just after the last mill stand, or at an intermediate location in the mill. To realize an improvement, the predicted settings, such as roll positions and roll forces, are made using models that are determined[12] prior to the entry of the strip to the roll gap. The strip was then rolled using these models with an improved and more rapid response. This method is sometimes denoted as the absolute gauge (AG) portion of the overall automatic gauge control, and has been successfully used in earlier applications.

[11] The mill stretch curve is similar to the curve of Fig. 4.9, except it is nonlinear and based on actual measurements.

[12] To the extent practical, models are refined using data obtained during previous processing.

4.3.1.3 Monitor Gauge Control

Irrespective of the previous efforts at improvement, there was still a deviation in the desired thickness of the final product, mostly due to the lack of an immediate measurement. What has been used, however, is an X-ray measurement mounted reasonably close to the mill exit. This results in the absolute measurement of the strip thickness, which can be used to provide a correction to the control at each mill stand. The function that develops and distributes the correction signals is denoted as monitor gauge control (MAGC), which is part of the overall AGC.

In general, this function consists of two parts. The first part is a fast responding proportional–integral (PI) controller that adjusts, at the last mill stand, the thickness of the head end of the strip between the last stand and the X-ray gauge. The second part is a much slower PI controller that adjusts the thickness of the remainder of the strip. The adjustment is spread over the individual stands to distribute the correction so that a few individual stands are not overloaded. The Smith predictor function is applied where needed to maintain the correct timing of the adjustments.[13] Overall, in an earlier application this approach has been reasonably effective in reducing the error in the final product.

4.3.1.4 Head End Threading

Just before threading, the mill is set up using models to determine initial roll gap settings and predicted roll forces. Using these models, the settings for stand 1 are made and the strip is processed using the AG function as described previously. Data on the processing at stand 1 are taken for use in feedforward control in downstream stands, and the head end of the strip is advanced to stand 2. At stand 2, the roll force measurement is evaluated to determine the trace of the hardness of the strip. Data are updated and additional data are taken to adjust for changes in hardness and other characteristics for processing in downstream stands. The data are passed by feedforward techniques to adjust downstream control using AG techniques. This similar method is used at subsequent stands to improve the overall strip characteristics at the exit of the last stand.

As the head of the strip reaches the X-ray gauge, the MAGC is performed to further improve the strip characteristics. Additionally, data on this strip are passed to the set-up function for the next strip to improve estimations of settings for the next strip. This was useful for processing a particular batch of similar product.

---

[13] More detail is available in [8]. Detail of the Smith predictor feedback technique is available in Appendix Section "Smith Predictor Characteristics".

4.3.2 *Mass Flow Techniques*

The need for improved mass flow techniques arose from the need to improve the responses of the automatic gauge control. This was especially so wherein screw-downs were replaced by hydraulic cylinders, and the angle of the looper was done by low inertia torque motors or other equivalent methods. These changes improved the responses of the looper-based system to various perturbations and disturbances. While some improvements were realized, there remained the slower response of the upstream stand drive to a change in the strip speed based on a change in the looper angle. Some examples are as follows.

4.3.2.1 Mass Flow Techniques, Example 1

An initial improvement was developed using a mass flow method wherein a change in the position of the cylinder (Fig. 4.6) of the downstream mill stand could be used to adjust the strip speed of the adjacent upstream mill stand. This was based on the conservation of mass flow through the downstream stand, assuming the strip width is unchanged. For example, in the downstream stand, as in (4.18), an increase in the exit strip thickness ($h_{\text{out},i+1}$) caused by a decrease in the cylinder position (Fig. 4.6) would result in an increase in the entry strip speed ($v_{\text{in},i+1}$),

$$h_{\text{in},i+1} v_{\text{in},i+1} = h_{\text{out},i+1} v_{\text{out},i+1}. \tag{4.18}$$

Since the response of the upstream drive system is a bit slow, the increase in $v_{\text{in},i+1}$ would reduce the length of strip between the two mill stands and thus reduce the looper angle, with a change in the mass flow. The response (Fig. 4.10) was to add a proportional–integral (PI) trim on the upstream stand drive strip speed based on the position of the hydraulic cylinder of the downstream stand.

This initial approach offered some improvements, and was partially successful. However, as noted earlier, the cylinder position in the downstream stand also could be affected by the strip hardness, the uncertainties in the entry gauge, plus others; and further, the effects of the transport time for the validity of the correction signal would need to be considered.

Additionally, the system is based on linearized models at a particular operating point. Success over the full range of operation required updates of the model and control as needed to assure stability and performance for each strip being processed.

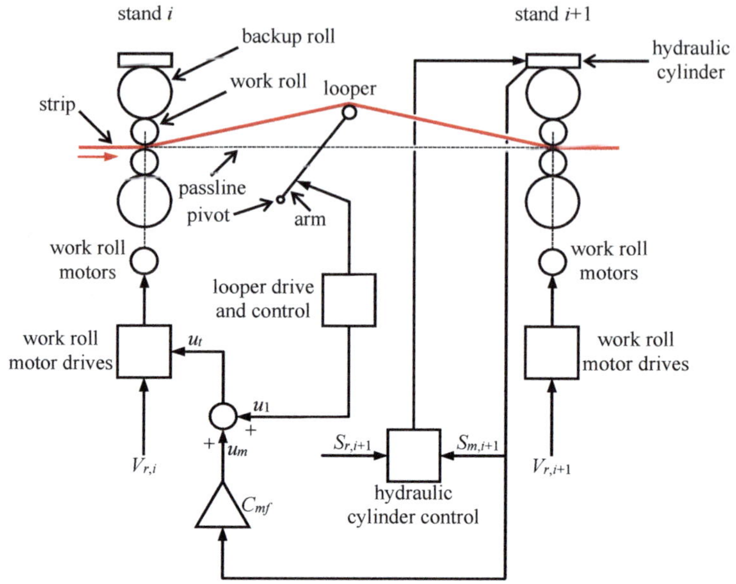

Fig. 4.10 Functional diagram[14] of earlier conventional mass flow control. u_t: trim of mill speed, stand i*[15]; u_l: looper control for trim of mill speed, stand i*; $V_{r,i}$: mill speed ref, stand i; $V_{r,i+1}$: mill speed ref, stand $i + 1$; $S_{m,i+1}$: position of hydraulic cylinder, stand $i + 1$[16]; $S_{r,i+1}$: hydraulic cylinder position ref, stand $i + 1$; C_{mf}: proportional–integral control for trim of mill speed, stand i; u_m: output of C_{mf}*; * trims are deviations from set-up values

4.3.2.2 Mass Flow Techniques, Example 2

An improvement in Example 1 takes into account two additional disturbances that can change the entry strip speed of the downstream mill stand. By doing this, a constant mass flow is better realized than in Example 1. The estimates of the three disturbances at the downstream stand are the sensitivities in the entry gauge, in the roll gap, and in the hardness of the strip. Of these three, of particular interest is the hardness. How this was done is described in the following.

The main idea is to improve the trim on the control of the main drive strip speed of the upstream stand to reduce fluctuations in the mass flow between the upstream and downstream stands. This was implemented by using a trim based on (1) the estimate

[14] This diagram supplements earlier diagrams for the looper and AGC controls.

[15] The position of the hydraulic cylinder (noted as S_m in Figs. 4.6 and 4.8) is controlled by AGC as in Fig. 4.8; e.g., in stand $i + 1$, a decrease in cylinder position causes an increase in the output strip thickness and thus an increase in the input strip speed, with a corresponding decrease in the looper angle.

[16] The work roll drives process u_t such that an increase in u_t causes an increase in mill stand i strip speed and thus an increase in the looper angle. Similarly, the drives for trims u_l and u_m are each configured such that an increase in the trim will increase the value of u_t; e.g., a decrease in $S_{m,i+1}$ would increase u_m, and a decrease in the looper angle would increase u_l.

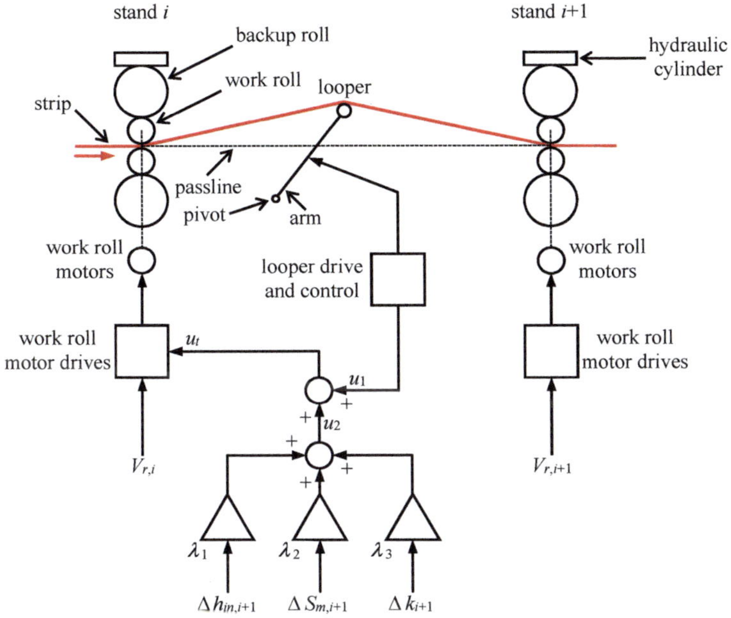

Fig. 4.11 Functional diagram of improved conventional mass flow control. u_t: trim of mill speed, stand i*[17]; u_1: looper control for trim of mill speed, stand i*; u_2: trim of mill speed per (4.19)*; $V_{r,i}$: mill speed ref, stand i; $V_{r,i+1}$: mill speed ref, stand $i+1$; k_{i+1}: input strip hardness, stand $i+1$**; $h_{in,i+1}$: input strip thickness, stand $i+1$**; $S_{m,i+1}$: position of hydraulic cylinder, stand $i+1$**; λ_1, λ_2, λ_3 are factors (4.19); * trims are deviations from set-up values; ** estimated

of the entry gauge at the downstream stand, (2) the estimated roll gap, and (3) the estimate of hardness at the downstream stand. The ultimate results are as in (4.19) and depicted in Fig. 4.11:

$$\Delta v_{\text{in},i} = \lambda_1 \Delta h_{\text{in},i+1} + \lambda_2 \Delta S_{\text{m},i+1} + \lambda_3 \Delta k_{i+1}. \tag{4.19}$$

In (4.19):

- $\Delta v_{\text{in},i}$ is the trim on the input to the control of the upstream stand strip speed;
- $\Delta h_{\text{in},i+1}$ is the estimated feedforward sensitivity of the strip thickness input at the downstream stand;
- $\Delta S_{\text{m},i+1}$ is the sensitivity in the gap of the downstream stand; and
- Δk_{i+1} is the estimated feedforward entry hardness sensitivity, using the estimated strip hardness at the upstream stand, where it is updated using related data for estimating the entry hardness at the downstream stand.

[17] The work roll drives process u_t such that an increase in u_t causes an increase in mill stand i strip speed and thus an increase in the looper angle.

The λ factors are functions determined by characteristics of the process model that can affect associated variables; for example, changes in the mill modulus M due to the changing characteristics of the mill stand. This can affect the mill stretch curve, as well as other changes that often occur during actual operation, and can also affect other mill models. These models are usually in the process computers (level 2) which interact with the mill controllers. Also, it is important to note that disturbances, measurement noise, undesirable interactions between variables, should be avoided in the functions between mill controllers or between controllers and process computers. Thus, in this application, the updating of hardness functions, and other applicable issues, is mostly via Kalman filters[18] to assure usable transmissions of data and accurate tracking between various variables.

This example as described briefly in [15] has been in use earlier for many years with good success, especially with large variations in hardness. Some data as depicted in [15] shows a reasonable reduction in undesirable deviations in looper angle, roll gap change, and mass flow trim, when compared to Example 1. However, it is not clear to what extent the more advanced methods, as referenced in [15] and described in [16], were applied during the entire time in use at and beyond the initial start-up.

Additionally, as noted in [16], the application of extended Kalman filters requires the design engineers and start-up personnel to develop and tune certain matrices that are essential for a desired high performance. This would require considerable knowledge and experience with the actual process, plus on-line tuning which assumes familiarity with more advanced control techniques. This could be less desirable, and should be considered in the development of various control methods.

4.3.2.3 Mass Flow Techniques, Example 3

In the previous examples each control loop had to be individually designed, with interactions between loops also being similarly considered individually. However, in using the multivariable concept, the design becomes easier, with the tuning being simpler and more intuitive. A major reason for this is that the multivariable controller inherently considers all measurements simultaneously using prior data so that undesirable interactions are considerably reduced. In this particular application as depicted functionally in Fig. 4.12, the multivariable control has been reasonably effective in realizing its main objective, i.e., to produce effective trims on the upstream drive speed control and on the downstream hydraulic cylinder control. An overall improvement in performance is realized, with lesser complications for design and start-up personnel. The following overview briefly describes this multivariable application [18]. More detail is available in Appendix Sections "Brief Overview of Kalman Filter as a State Estimator" and "Additional Detail of the Multivariable Control".

In this case the control is linear with a linearized nonlinear model. The model is depicted in state-space form as:

---

[18] Some additional detail on the derivation of (4.19) is in Appendix Sections "Derivation of Mass Flow Techniques", "Brief Overview of Kalman Filter as a State Estimator", and [16, 17].

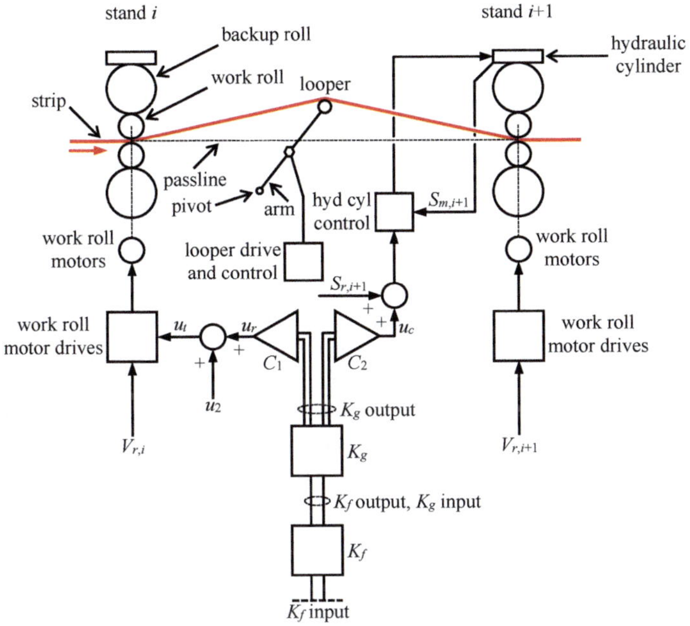

Fig. 4.12 Functional diagram of control using multivariable techniques. u_t: trim of mill speed, stand i (Fig. 4.11)*; u_2: trim of mill speed, stand i (Fig. 4.11)*; u_r: trim of mill speed, stand i*; u_c: trim of cylinder position, stand $i+1$*; $V_{r,i}$: mill speed ref, stand i; $V_{r,i+1}$: mill speed ref, stand $i+1$; $S_{r,i+1}$: hydraulic cylinder position ref, stand $i+1$; $S_{m,i+1}$: position of hydraulic cylinder, stand $i+1$; C_1: proportional–integral–derivative (PID) trim for mill speed, stand i; C_2: PID trim for hydraulic cylinder position, stand $i+1$; K_g: gain matrix provides inputs to C_1 and C_2; K_f: Kalman filter provides state variables for K_g, e.g., looper angle; * trims are deviations from set-up values

$$\dot{x} = Ax + Bu + w_d, \quad x(0) = x_0, \tag{4.20}$$

and

$$y = Cx + Du + w_n, \tag{4.21}$$

where A, B, C, and D are determined by the linearization of the nonlinear system at the operating point; where x, y, and u are the system states, measurements, and inputs; and where w_d and w_n represent the state disturbances and measurement noise.

An observer that determines the actual states being used is the Kalman filter which is an optimal linear observer.

The states at the output of the Kalman filter are:

- x_1: looper angle;
- x_2: cylinder position, downstream stand;
- x_3: drive speed, upstream stand;
- x_4: entry strip thickness disturbance, downstream stand;

- x_5: entry strip hardness disturbance, downstream stand;
- x_6: mass flow disturbance.

The inputs to the Kalman filter are based on using the present linearized model. Measurements are predicted using this model. Actual measurements then are taken and compared with those predicted, and a correction is applied by the Kalman filter to realize the states for use in determination of the control. The output of the Kalman filter is applied to a gain matrix. The gain matrix applies linear quadratic regulator (LQR) methods[19] to determine references for the proportional–integral–derivative (PID) functions for the trim of the upstream speed controller and the trim of the downstream gap via the cylinder position controller.

In this instance a separate output was provided for each of the proportional, integral, and derivative functions for each trim. It was considered that this might help during design and tuning functions, as there would be some similarity to conventional control. An additional feature of this technique is that the elements of the gain matrix are updated for each strip being processed, plus the updating of the LQR cost factors to retain good dynamic performance over a wide range of products.

The previously described method was applied to an actual mill with reasonable success. The improvement in gauge was not major, but was consistent over a large number of coils being processed, including coils of various characteristics. Thus, such an improvement was useful, considering that the quality and yield of the overall product were maintained for an extended period.

Additionally, the multivariable method has resulted in very low excursions in the strip tension. This has occurred due to a reduction in the deviations in the looper angle, a rapid response of the looper drive torque control to disturbances, plus a reasonably accurate compensation of the inertia of the looper system mechanics. Moreover, when compared to earlier conventional control, the peak of the looper angle transient that usually occurs during the initial contact with the strip during threading was reduced. Also reduced was the time for the angle to return near to its steady-state value, which is important in realizing better threading performance.[20]

Overall, the usage of multivariable control was determined to be useful, with advantages over single-input–single-output systems. The work as previously described provided a basis for future investigations that could lead to further improvements. A possible example is the use of nonlinear process models and multivariable nonlinear control. This could reduce some of the issues that must be addressed in a linearized system to assure good performance over the entire range of operation, and also considering possible abnormal events. Some of these methods are described in Chap. 5.

---

[19] An example of the LQR method is in [19].

[20] More detail regarding improvement in performance of an actual process is available in [18].

4.3.3 Strip Thickness Control Throughout the Mill

The application of what was previously described is coordinated throughout the mill to realize an improvement in the final product. Some of the more significant areas are as follows.

As noted previously, the position of the looper plays an important part in reducing deviations in the mass flow between adjacent mill stands, as well as maintaining a nearly constant strip tension. Thus, holding the looper position to have very little deviation is crucial. This is done by the trimming of the upstream stand motor drive for the strip speed, and the trimming of the downstream stand cylinder position for the trimming of the gap.

However, there are additional disturbances that also can affect the looper position that occurs due to the change of the downstream stand exit gauge. Some of these (as noted previously in Sect. 4.3.2.2) are the estimated downstream stand entry gauge, the estimated downstream stand entry hardness, and others such as those described in Sect. 4.3.1.1. These can occur not only during threading, but also during processing of the entire length of strip. Additionally, to assure that the trims on a certain stand motor drive (e.g., stand i), and the control for the associated looper position (between stands i and $i + 1$), do not significantly affect the further upstream looper positions, appropriate trims are applied to the previous upstream stands.

As also described previously in Sect. 4.3.2.3, the use of multivariable control can simplify the control of the trims of the upstream speed and downstream gap.

4.3.3.1 Modeling Issues

Having models that reasonably represent the actual process is necessary for high-quality performance during set ups for threading as well during actual operations. This would include the flexibility in modeling to handle a wide variety of possible conditions. Some examples are a reduction in mass flow disturbances, handling well a wide variation in strip and mill characteristics such as temperature variations, the capability to retain good performance during processing of special products such as thinner materials, plus others.

Unfortunately, because of the more complex and hostile mill environment, even the most reliable measurements are unable to provide the stringent accuracies required to realize the requirements desired in the final product. Thus a learning system is needed to attempt to keep the models as close to the actual process as is possible and reasonable. In some instances, special techniques have been developed using mathematical techniques to significantly reduce the variables' uncertainty.

Additionally, the models often require updating to handle the change in the strip characteristics during processing in the downstream stands. An example is the change in the strip hardness realized in the upstream stands, so that, if possible, some changes in the downstream stands can be realized to improve the output. In certain instances, techniques using extended Kalman filters are used to estimate the hardness, so that downstream stands can offer correction. Further, modern systems include more sophisticated techniques to handle mill and product variations. More specific details about these and more advanced methods are usually available via mill and control system providers.

4.3.3.2 Use of Feed Forward

In the case of the entire mill, the application of feed forward of strip characteristics is quite useful. For example, the estimates of hardness for a section of the strip are tracked during its complete processing from stand to stand. During this procedure, in a section of the strip, the validity of the interactions of the estimated hardness with other interacting variables and associated functions is assured using techniques based on the use of Kalman filters. This coordination allows an earlier estimation of material characteristics using feedforward methods before a section of strip reaches the next downstream stand. This improves response times, and is especially effective during threading.[21]

4.3.3.3 Adjustment of Drive Loading and Gauge Distribution

Further, sometimes during processing, an estimation of downstream stand characteristics is done at upstream stands to perform an adjustment of the output gauge at certain upstream stands. This results in a better distribution of the actual downstream stand drive loadings and roll forces required to process the strip. This also ultimately results in better gauge performance in downstream stands as lesser and faster changes in the hydraulic cylinder positions are realized to achieve changes in the downstream gaps. An example is depicted in Fig. 4.13.

Additionally, the loading of the drives is monitored[22] to determine the potential of overloading one or more drives, or the potential for a drive to be in a condition that restricts its capability. In this case the gauge out of certain upstream stands is adjusted to redistribute the load to avoid any overloading that would deteriorate the drives' capability, and thus avoid undesirable conditions such as deterioration in strip flatness, or the restriction in speed that could lead to effects of higher strip temperatures.

However, irrespective of these methods, usually there are remaining deviations in the gauge at the exit of the last stand. These errors are detected by an X-ray gauge located after the last stand to reduce the remaining gauge deviations. To account for the large transport delay from the last stand to the X-ray gauge, Smith predictor

[21] Additional detail is in [20].

[22] The monitoring of drive current can be insufficient to determine the drive torque; e.g., in most systems the torque also depends on the machine flux, which could be weakened during various operations such as zooming of the mill. In most cases, especially with modern variable speed drives, it is best to coordinate with the drive supplier to receive an appropriate signal of the drive's capability. More detail is in Chap. 6.

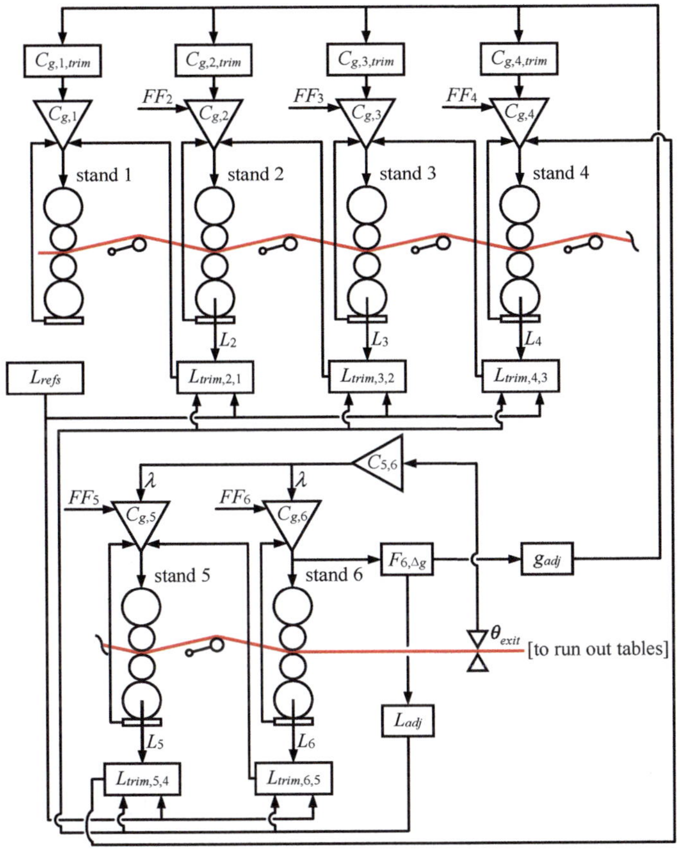

Fig. 4.13 Functional description of drive loading and gauge distribution.[23] θ_{exit}: mill exit temperature; $C_{\text{g},1}$–$C_{\text{g},6}$: gauge controls, stands 1–6; $C_{\text{g},1,\text{trim}}$–$C_{\text{g},4,\text{trim}}$: gauge control trims, stands 1–4; $C_{5,6}$: gauge control trims, stands 5–6, Smith predictor; λ: load sharing, stands 5–6; L_{refs}: force trim refs, stands 2–6; L_{adj}: force trim refs adjust, stands 2–6; L_2–L_6: drive loads, stands 2–6; $L_{\text{trim},2,1}$–$L_{\text{trim},6,5}$: drive load trims for load distribution, stands 2–6; g_{adj}: adjust for redistribution of gauges, stands 2–4; $F_{6,\Delta\text{g}}$: estimations of L_{adj} and g_{adj}; FF_2–FF_6: feed forward, stands 2–6

feedback is applied to use the outputs from the X-ray gauge to compensate for the delay by correcting for model inaccuracies at the last two stands.[24] The Smith feedback is used at both stands together to share any additional needed loading, and thus avoid overloading a single stand.

---

[23] Copyright Association for Iron and Steel Technology 2009, Fig. 4.13. Portions of this figure are adapted and cosmetically reprinted, by permission, from [21].

[24] Two exit stands are typical. In some cases there is a different number of exit stands which are treated similarly to the two exit stands.

4.3.3.4 Eccentricity Control

Eccentricity is a general term that refers to any condition that would result in an irregularity in the mill rolls or roll bearings. Such irregularities can cause cyclic deviations in the mill load, which results in an undesirable change in the strip thickness. Eccentricity effects are mostly due to the backup rolls since they are greater in diameter than the work rolls. Some examples are (1) out-of-roundness of a roll, (2) eccentricity of backup roll journals with respect to the roll body, (3) non-uniformity of rollers in the roll bearings, plus others. Additionally, due to the cyclic deviations, and since the upper backup roll and the lower backup roll are usually of different diameters and out-of-round, the result is a modulation effect of much lower frequency.[25] However, the work rolls can also contribute to the eccentricity although less severely.

Also, as noted previously (Chap. 3), the difficulties are intensified since the resulting interaction with the AGC can be misunderstood by the BISRA feedback as a change in the characteristics of the strip. This then results in a move in the wrong direction by the cylinder controlling the gap. Thus, as might be expected, the design of control to mitigate the effects of eccentricity poses a significant challenge. For example, it is crucial that the eccentricity control works well with the automatic gauge control system and does not degrade its performance. Additionally, other issues of concern are as follows:

- the eccentricity characteristics are strongly based on the speed of the mill; this is especially applicable in a hot mill during threading and zooming;
- there can be slight differences in backup roll diameters, such as the changes in their diameters during changes in temperature and roll wear;
- some disturbances, such as the effects of a keyway, should be addressed;
- measurements can be less accurate or unavailable; e.g., strip thickness, stick–slip or hysteresis of the mill stand, and others;
- be useful in a practical sense to the design and start-up personnel, most of whom are less familiar with advanced control techniques; and
- others, peculiar to the particular application.

As noted, the development of a reasonably successful control of eccentricity can be quite challenging. In response to this challenge, over the years there have been a wide array of technical papers, patents, etc., in this area. To provide some initial insight in this area, what follows are some earlier examples of basic ideas.[26] More advanced techniques are described in Chap. 5.

4.3.3.5 Eccentricity Control, Example 1

In this example [22] an earlier method of active eccentricity compensation generally consisted of two parts. The first part was the identification of the amplitude and

[25] An example of eccentricity is depicted in Appendix Section "Example of Eccentricity".

[26] Additional examples are available in [23].

phase of periodic disturbances. The second part was to reject the induced periodic vibrations, so that the overall response was to determine and respond to the total roll eccentricity, based on the determination of the total axial deviation of the backup roll.

Two methods of estimating deviations in the exit strip thickness could be applied. The first uses the relationship (4.16) for the deviation around an operating point:

$$\Delta h_{\mathrm{out,e}} = \frac{\Delta F_{\mathrm{roll}}}{M} - \Delta S_{\mathrm{m}} + e, \tag{4.22}$$

where $h_{\mathrm{out,e}}$ is the estimated output thickness, F_{roll} is the roll force, M is the mill modulus, S_{m} is the position of the gap adjust cylinder, and e is the eccentricity, i.e., the elastic deformation of the mill stand. A second method is:

$$\Delta h_{\mathrm{out,e}} = \frac{\Delta F_{\mathrm{roll}}}{k_{\mathrm{p}}}, \tag{4.23}$$

where k_{p} is the plastic deformation constant of the strip. Unfortunately, both of these methods rely heavily on M and k_{p} being quite accurate for a variety of products and operating situations, which was not the case. Additionally, the measurements were quite sensitive to measurement noise, and further, the control would interfere with the stand AGC, which was undesirable. Thus overall, the method was only partially successful.

4.3.3.6 Eccentricity Control, Example 2

In this example the deviations in the roll force caused by eccentricities were used as trims to the thickness controller [23]. This was implemented by determining the eccentricities with the mill stopped, without strip, and the rolls closed. Measurements of the roll force were then taken at full rotations of the upper and lower backup rolls, to determine the deviations in the roll force caused by the eccentricities as a function of the angles of each roll.

These functions (approximated as sinusoidal) were added to the control of the mill thickness to compensate for the eccentricates as in Fig. 4.14, as described in Fig. 4.8. In Fig. 4.14, $S_{\mathrm{e,u}}$ and $S_{\mathrm{e,l}}$ represent the trims on the cylinder position based on the angular positions of the upper and lower backup rolls as in (4.24) and (4.25):

$$S_{\mathrm{e,u}} = \left(\frac{F_{\mathrm{A,u}}}{M}\right)\sin(\theta_{\mathrm{u}} + \alpha_{\mathrm{u}}), \tag{4.24}$$

$$S_{\mathrm{e,l}} = \left(\frac{F_{\mathrm{A,l}}}{M}\right)\sin(\theta_{\mathrm{l}} + \alpha_{\mathrm{l}}). \tag{4.25}$$

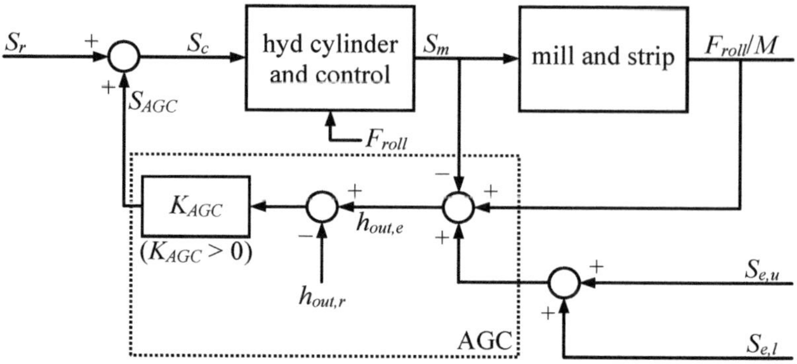

Fig. 4.14 Diagram of the basic AGC (Fig. 4.8) with trims for eccentricities

In (4.24), $F_{\mathrm{A,u}}$, θ_{u}, and α_{u} are the characteristics of the roll force of the upper backup roll determined during the initial measurements. $F_{\mathrm{A,u}}$ is the peak roll force due to roll eccentricity; M is the mill modulus; θ_{u} is the angle between selected zero position and the position corresponding to the average rolling load; α_{u} is the angular position based on θ_{u}; and similarly for the lower backup roll (4.25).

This earlier method was somewhat effective, but with some drawbacks. Particularly, the technique was not applicable for a mill stand with a single drive for both upper and lower work rolls, which was typical of certain mills, usually on the first stand. Additionally, the method is based entirely on a sinusoidal eccentricity, which is usually not the case, as the effects of higher frequencies also should be considered; the lesser effect of the work rolls on the eccentricity was not addressed; considerable work was needed to set up the positions on the backup rolls to establish the characteristics of (4.24) and (4.25); and as in Example 1, the measurements of M and others are sensitive to noise and uncertainties. A somewhat improved later method is described in Example 3.[27]

4.3.3.7 Eccentricity Control, Example 3

In this example two of the major weaknesses of Example 2 are addressed. These are (1) the requirement for separate drives for the upper and lower work rolls, and (2) the time required for implementing the method used for determining the roll eccentricity, so that the separate measurements required for each backup roll are unnecessary.

In this method, as in Example 2, the mill is stopped, the rolls are brought together, except that they are just touching without strip or load (i.e., a "kiss" condition). The rolls are then rotated to develop a beat cycle that represents the eccentricity. In this case the force signals (of the fundamental frequency) are separated for each roll by one beat cycle time. The angular positions are measured by pulse generators

[27] Several other examples are available in [23] and the references cited therein, and in other similar references.

connected to the rolls, so that the peaks of the sinusoids, and thus the individual signals of each roll can be determined. The signals are then applied as in Example 2. In event there are higher frequencies present possibly due to harmonics, a similar idea can be used, but with a bit more complicated calculation.

4.3.3.8 Profile and Flatness Control

The quality of the final product includes the acceptability of the strip's profile and flatness. The crown k_c of the profile is understood (4.26) as the center thickness d_3, with respect to thicknesses d_1 and d_2 as measured at a specified distance from each side of the strip, as depicted in Fig. 4.15. The strip is understood to be flat if it can lie flat on a flat surface under the influence of only gravity.[28]

$$k_c = d_3 - \frac{(d_1 + d_2)}{2}. \tag{4.26}$$

In many mills the upper and lower work rolls are deliberately shaped to produce a system that, by shifting axially, a continuously variable crown (CVC) can be realized. Examples are depicted in Fig. 4.16, where shifting can produce a positive or negative crown in the strip. However, crown also can occur from other conditions. Some of these are roll wear, roll bending, change in temperature, plus others, all of which must be considered in the control.

In the case of flatness, undesirable results can occur if there is excessive crown, either in the center or at the edges of the strip. If crown is at the edges, there will be waviness in the strip with the edges of the strip turned upward; if at the center, there also will be waviness in the strip, except with a buckle in the center. In both cases, these conditions are caused by differences in lateral elongations resulting from excessive crown; e.g., excessive positive crown can result in edge waviness, and excessive negative crown resulting in center buckling. This elongation is generally described in (4.27) as:

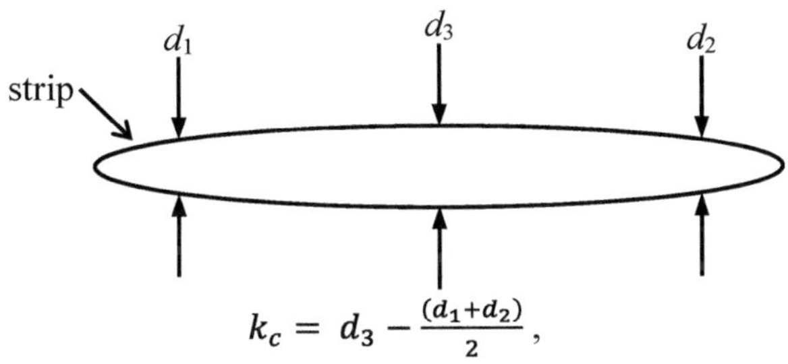

Fig. 4.15 Functional diagram, strip cross section showing determination of crown k_c

---

[28] A more formal definition of flatness is the geometric departure from a reference plane when the strip is not under tension and at ambient temperature [24].

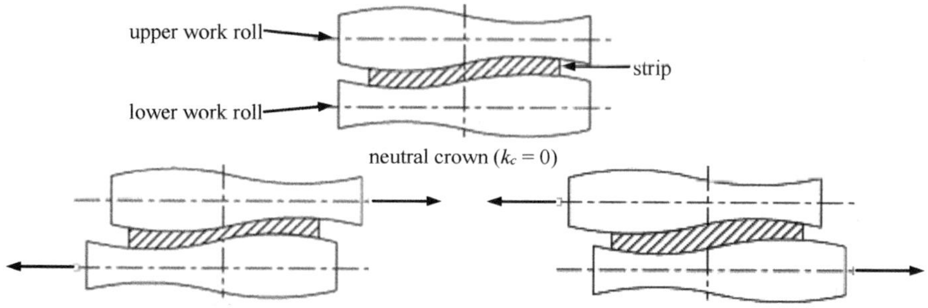

Fig. 4.16 Examples of crowns with CVC[29]

$$\delta = \frac{k_{c,\text{in}}}{h_{\text{in}}} - \frac{k_{c,\text{out}}}{h_{\text{out}}}, \tag{4.27}$$

where $\frac{k_{c,\text{in}}}{h_{\text{in}}}$ and $\frac{k_{c,\text{out}}}{h_{\text{out}}}$ represent the elongations at the input and output of a stand.

Generally, most strips can tolerate some crown if they are within a certain range [23] of flatness as in (4.28), typical for many facilities:

$$-80\left(\frac{h_{\text{out}}}{W}\right)^{1.86} < \delta < -40\left(\frac{h_{\text{out}}}{W}\right)^{1.86}, \tag{4.28}$$

where W is the strip width. However, a certain amount of width and crown is needed to keep a required profile if the strip is to be used for cold rolling, galvanizing, or certain other applications. Thus if an acceptable crown is realized, usually an acceptable flatness also will be obtained.

The crown is strongly affected by the work roll bending jacks and the main cylinder for control of the roll gap. Due to the significance of the profile and flatness, there have been many methods of control implemented and proposed to improve the quality of the crown in the final product. What follows is a brief functional description of an earlier method as more fully described in [23] to give some basic insight into the issues involved. More detailed and advanced methods are described in applicable references, e.g., [25–29].

4.3.3.9 Profile and Flatness Control, Example

The modeling of the mill set-up involves determination of the setting of the actuators, such as the work roll bending jacks, to realize the desired crown and flatness of the head end of the strip during threading and subsequent processing. However, in certain instances, the roll bending functions may be insufficient to meet the requirements

[29] Adapted from [23]. Copyright 1993 Marcel-Dekker, courtesy of Taylor & Francis (all rights reserved).

of the specifications for the particular application. Where this is the case, it usually requires the use of continuously variable crown (CVC) rolls, as previous described.[30] The desired model is then determined and entered into the mill computer system (level 2).[31] It is important also to review the possible interactions with the strip thickness control systems to assure good performance and reduce the possibility of undesirable interactions. The set-ups for conditions are as follows, assuming the mass flow is the same throughout the mill; for achieving desirable flatness, the variable λ_n is to be the same at the exit of each stand n of the mill,[32] where

$$\lambda_n = k_{c,n}/h_n + \delta_n, \tag{4.29}$$

i.e., $\lambda_1 = \lambda_2 = \lambda_3 = \ldots \lambda_n$, where $k_{c,n}$ is the strip crown, h_n is the strip thickness, and δ_n is the dead band in the flatness; the roll force F_n at stand n should conform reasonably closely to (4.30):

$$F_n = r_n w_n \left[R_n (h_{n-1} - h_n) \right]^{0.5}, \tag{4.30}$$

where r_n is the strip hardness, w_n is the strip width, R_n is the flattened roll radius, and h_{n-1} and h_n are the strip thicknesses at the entry and exit of stand n.

The controller,[33] supported by the model, consists of two parts, the profile control and the flatness control as described in [23], and is depicted functionally as an overview in Fig. 4.17. In this example the upper three stands of a six-stand mill were used for control of the strip profile, while the lower three stands were used for control of the flatness.

The objective of the profile control is to closely realize the desired relative strip crown before the strip gets to stand 4. Thus the strip crown just after stand 3, which is measured by the sensor $S_{p,1}$, should be close to the desired crown, i.e.:

$$k_{c,3} = k_{c,6} \frac{h_3}{h_6}. \tag{4.31}$$

The difference $\Delta k_{c,3}$ is used to adjust the difference in the roll bending force $\Delta F_{b,3}$ of stand 3 as

$$\Delta F_{b,3} = b_{br} \Delta k_{c,3}, \tag{4.32}$$

---

[30] The CVC rolls are initially installed based on the determination of fixed positions and do not move during the actual rolling process.

[31] Some detailed calculations are described in [27].

[32] In this application, the stand draft rolling loads in the downstream stands would be somewhat decreased below the load for maximum production, while the load at the upstream stands would be somewhat increased.

[33] The description of this application is an overview of some of the main ideas.

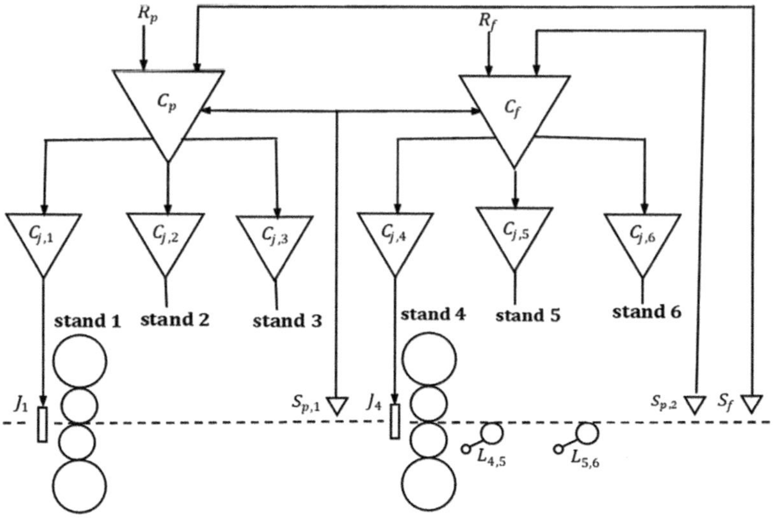

Fig. 4.17 Functional earlier example of profile and flatness control.[34] R_p: profile reference; R_f: flatness reference; C_p: profile control; C_f: flatness control; $C_{j,1}$: work roll bending jacks control, stand 1; $C_{j,2}$–$C_{j,6}$: work roll bending jacks control, stands 2–6; J_1: work roll bending jacks, stand 1*; J_4: work roll bending jacks, stand 4*; $S_{p,1}$: profile sensor 1; $S_{p,2}$: profile sensor 2; S_f: flatness sensor; $L_{4,5}$: shape sensing on looper, stands 4–5; $L_{5,6}$: shape sensing on looper, stands 5–6; * typical for stands 2, 3 and 5, 6. It was considered [23] that the sensors would operate properly in the environment. This needs to be addressed in an actual application

where b_{br} is the bending rate. These adjustments are intended to assure reasonably good conformance to the requirements of (4.29).

A main function of the flatness control in the downstream stands is to develop a high-quality flatness in the strip exiting the last stand. This would result in a nearly constant crown-to-thickness ratio which would be evident in the following relationships, which indicates a high-quality product,

$$\frac{k_{c,3}}{h_3} = \frac{k_{c,4}}{h_4} = \frac{k_{c,5}}{h_5} = \frac{k_{c,6}}{h_6}. \tag{4.33}$$

In Fig. 4.17 the shape sensing loopers[35] are designed with sensors across the strip width to measure the strip flatness. The measurements are used to trim the roll bending forces on the associated mill stands. The two sensors for profile and flatness provide signals for adjustments in the final product plus signals to update the computer models for upcoming strips.

[34] Adapted from [23]. Copyright 1993 Marcel-Dekker, courtesy of Taylor & Francis (all rights reserved).

[35] More detail on shape sensing loopers is in [23] and the cited references therein.

Many of the concepts as previously presented in this example were expanded upon later as a basis for further improvement. These are described more fully in the previously cited references.

4.4 Strip Temperature Control

Temperature variations in the strip are important as they can significantly affect the physical properties of the strip, especially the strip hardness and thus ultimately the strip gauge. Some of these variations that occur just prior to the mill entry are skid chills that have occurred in the cooling furnace, temperature reductions due to longer time on the delay tables, temperature run-up due to entry from the coil box (if used in the mill),[36] effects from high-pressure sprays to remove scale prior to the mill entry, plus others. Each of these can affect the measurement of the temperature at the mill entry, even if a reliable temperature sensor is available. If a reliable sensor is unavailable, the temperature is estimated using models and measured strip characteristics at the entry stand.[37]

The strip temperatures during rolling are controlled by the strip speed and water sprays at interstand locations. The requirements for spray water flows at the entry stands are mostly heavier than at the other stands. This is because strip temperatures at the entry end are generally hotter, which are more likely to contribute to the formation of scale that is highly undesirable during the rolling process. Also, allowing rapid changes in the temperature along the length of the strip can cause rapid changes in the strip hardness, which puts a greater stress on the gauge control function.

Unfortunately, at these interstand locations reliable measurements are not available mostly due to the water sprays and steam. Thus temperatures are estimated using the measurement at the mill exit and models that have been previously verified to the extent practical, with on-line adjustments as needed. This measurement is reasonably reliable and is also used in control of the sprays for cooling on the run-out tables. Thus, the concept of the temperature controller relies heavily on this measurement. Figure 4.18 depicts a functional example of a typical cooling control system.

The general idea behind the temperature controller is to work backward from the reliable measurement at the strip exit to establish a desirable method of controlling the strip speed and the individual sprays. In an example, as described in [20] and depicted functionally in Fig. 4.18, a temperature measurement of θ_{exit} is made at the exit pyrometer. Then, based on this measurement and using the models, an estimate is made of what the temperature of a section of the strip at the exit of stand 5 would be to realize this measurement when that section arrives at the pyrometer. The adjustment

[36] In the coil box, the head and tail of the strip are interchanged due to the coiling and then uncoiling actions.

[37] Temperature is estimated using a change in yield stress by measurements of roll force and gap.

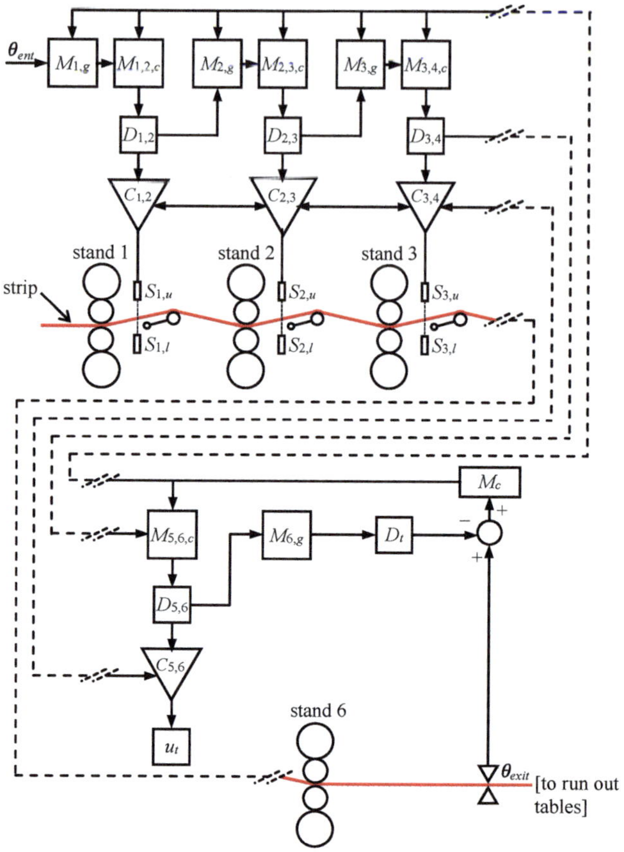

Fig. 4.18 Functional diagram of strip temperature control using multivariable techniques.[38] θ_{ent}: mill entry temperature; θ_{exit}: mill exit temperature; $M_{1,g}$: roll gap model, stand 1*; $M_{1,2,c}$: cooling model, stands 1, 2*; $D_{1,2}$: time delays to $C_{1,2}$ and $M_{2,g}$*; $C_{1,2}$: control for sprays between stands 1, 2*; $S_{1,u}$: upper water spray between stands 1, 2*; $S_{1,l}$: lower water spray between stands 1, 2*; D_t: time delay to temperature sensors (∇); M_c: model corrections; u_t: mill speed reference trim; * similar for stands 2–6, except that $C_{5,6}$ is the control for u_t

in the speed trim u_t is then computed[39] as needed to realize this temperature. Similarly, estimates of adjustments at other upstream stands are made, so that when the desired actuators are activated at the downstream stands, in accordance with the method of control, the upstream stands also are prepared in advance.

[38] Copyright 2005 Association for Iron and Steel Technology, Fig. 4.18. Portions of this figure are adapted and reprinted with cosmetic alterations, by permission, from [20].

[39] In the case as depicted in Fig. 4.18, it is assumed that there are no sprays at the exit of stand 5. If there were sprays in this location, they would be adjusted accordingly.

Some good results from application of this concept on an actual mill [20] are as follows. A more detailed treatment, including more advanced modeling and simulation, can be found in associated references, e.g., [16].

Mill characteristics	Controller performance
Product: Grade: 3300 Gauge: 8.1 mm Width: 1.53 m	Desired temperature: 860 °C
Stands: 7	Peak temperature: 869 °C[a]
Coil box: yes	Min temperature: 851 °C[a]
Sprays: after stands 1–4	
Processing time: 32 s	
Speed: ramp, 5–6.6 m/s	

[a] Measured at exit pyrometer

4.5 Concluding Comments

This chapter has presented the control of various areas of the tandem hot metal strip mill. The objective was to present some of the major issues and how several conventional methods have addressed them. Additionally, it presented some of the basic characteristics of the process plus several practical concepts to better understand the requirements for effective and useful results. It is felt that this would help to realize further improvements and advancements in these areas.

Appendix

Interaction Between Strip Tension Stress and Looper Angle

The following (from Chap. 2) shows the relationship between interstand strip tension stress and the looper angle that results in difficulties in establishing a desirable controller performance.

The interstand strip tension stress σ is modeled based on a relationship for Young's modulus wherein the strain depends on the length of the strip between adjacent mill stands:

$$\frac{d\sigma}{dt} \equiv \dot{\sigma} = \frac{E}{L_0}\left[\frac{dL(\theta(t))}{dt} + V_{\text{in},i+1} - V_{\text{out},i}\right], \quad \sigma(0) = \sigma_0, \tag{4.34}$$

where E is Young's modulus, L_0 is the distance between the centerlines of adjacent stands, $L(\theta(t))$ is the strip length between adjacent stands which depends on the position of the looper, $V_{\text{in},i+1}$ is the strip speed at the input to stand $i + 1$, and $V_{\text{out},i}$ is the strip speed at the output of stand i.

The angle of the looper position is expressed as:

$$\frac{d\theta}{dt} = \omega, \quad \theta(0) = \theta_0, \tag{4.35}$$

where θ is the angle of the looper arm, and ω is the angular velocity of the looper as determined using Newton's second law of motion:

$$\frac{d\omega}{dt} = \frac{1}{J_{\text{lpr}}}\left[M_{\text{lpr}} + M_{\text{fct}} + M_{\text{ld}}\right], \quad \omega(0) = 0, \tag{4.36}$$

where J_{lpr} is the moment of inertia of the looper, M_{lpr} is the torque applied at the looper, where M_{fct} is the frictional torque of the looper, and M_{ld} is the looper load torque as determined by the torques resulting from the strip tension force M_{ten}, the strip weight M_{swt}, the looper mass M_{lmas}, and the strip bending M_{bnd} at the looper roll:

$$M_{\text{ld}} = M_{\text{ten}} + M_{\text{swt}} + M_{\text{lmas}} + M_{\text{bnd}}. \tag{4.37}$$

More detail and some numerical examples are given in Chap. 2.

Derivation of Optimal Control

In support of Sect. 4.2.3.2 further improvement, Example 2, the following describes the use of the algebraic Riccati equation for optimality, and the addition of an integrator to realize a zero steady-state error in the strip tension and looper angle.

A revised state equation and the corresponding quadratic cost function are

$$\frac{dx_{\text{a}}}{dt} = A_{\text{a}}x_{\text{a}} + B_{\text{a}}u_{\text{a}}, \quad x_{\text{a}}(0) = x_{\text{a},0}, \tag{4.38}$$

and

$$J = \frac{1}{2}\int_0^{\infty} \left(x_{\text{a}}'Qx_{\text{a}} + u_{\text{a}}'Ru_{\text{a}}\right)dt, \tag{4.39}$$

where Q and R are the state and control weighting matrices, with Q positive semi-definite and R positive definite, and J is the performance index that is to be minimized with respect to u_{a}, where

$$x_{\mathrm{a}} = \begin{bmatrix} y_{\mathrm{r}} - y \\ \mathrm{d}x/\mathrm{d}t \end{bmatrix}, \quad \text{with } y_{\mathrm{r}} = \begin{bmatrix} \sigma_{\mathrm{r}} \ \theta_{\mathrm{r}} \end{bmatrix}', \tag{4.40}$$

and

$$u_{\mathrm{a}} = \frac{\mathrm{d}u}{\mathrm{d}t}. \tag{4.41}$$

The control law that minimizes J is (4.43), and is determined by the solution K of the algebraic Riccati Eq. (4.42), where

$$A_{\mathrm{a}}'K + KA_{\mathrm{a}} - KA_{\mathrm{a}}B_{\mathrm{a}}R^{-1}B_{\mathrm{a}}'K + Q = 0, \tag{4.42}$$

and

$$u_{\mathrm{a}} = -R^{-1}B_{\mathrm{a}}'Kx_{\mathrm{a}}. \tag{4.43}$$

The matrices A_{a} and B_{a} are determined such that the pair $(A_{\mathrm{a}}, B_{\mathrm{a}})$ is stabilizable[40] over the entire range of operation to assure a solution to (4.42). Additionally, the pair includes the matrices as in the following:

Taking

$$B_{\mathrm{a}}' = \begin{bmatrix} B_1' \ B' \end{bmatrix}, \tag{4.44}$$

and

$$u_{\mathrm{a}} = -R^{-1}\begin{bmatrix} B_1' \ B' \end{bmatrix}\begin{bmatrix} K_{11} & K_{12} \\ K_{21} & K_{22} \end{bmatrix}x_{\mathrm{a}}; \tag{4.45}$$

and taking

$$B_1' = \begin{bmatrix} 0 & 0 \\ 0 & 0 \end{bmatrix}, \tag{4.46}$$

then

$$u_{\mathrm{a}} = -R^{-1}B'\begin{bmatrix} K_{21} \ K_{22} \end{bmatrix}x_{\mathrm{a}}, \tag{4.47}$$

and

$$\frac{\mathrm{d}u}{\mathrm{d}t} = -R^{-1}B'\begin{bmatrix} K_{21} \ K_{22} \end{bmatrix}\begin{bmatrix} y_{\mathrm{r}} - y \\ \dfrac{\mathrm{d}x}{\mathrm{d}t} \end{bmatrix}x_{\mathrm{a}}, \quad u(t_0) = U(t_0), \tag{4.48}$$

---

[40] The definition of stabilizable, and additional detail on the use of the algebraic Riccati equation, are available in [30] and the references cited therein. Validation of stabilizability is by simulation.

and

$$u = -R^{-1}B'K_{21}\int_{t_0}^{t}(y_r - y)dt - R^{-1}B'K_{22}\int_{x(t_0)}^{x(t)}dx + U(t_0). \tag{4.49}$$

Brief Overview of H$_\infty$ *Methods*[41]

The H-infinity norm (4.50) is defined in a Hardy space (H_∞)[42] which includes all the proper, real, and rational stable transfer function matrices, and is

$$\|F(S)\|_\infty = \omega \sup \overline{\sigma}[F(j\omega)], \tag{4.50}$$

where $\overline{\sigma}$ is the largest singular value of F and ω is the (real) frequency.

The H_∞ *mixed sensitivity method* uses a linear state-space representation of the actual process, as

$$\frac{dx}{dt} = Ax + Bu, \quad x(0) = x_0, \tag{4.51}$$

$$y = Cx, \tag{4.52}$$

where $x = [\sigma\ \theta\ \omega\ V_R\ \beta]'$, $y = [\sigma\ \theta]'$, $u = [u_1\ u_2]'$, as noted in Fig. 4.5.

The symbol (′) indicates the matrix transpose, and A, B, and C are constants established by the linearization of the process model at an operating point. The mixed sensitivity method is applied to the process described in Fig. 4.19, wherein $C(s)$ and $P(s)$ represent the control and the plant, respectively, $W_1(s)$ and $W_2(s)$ are weights to be established, and s is the Laplace transform. The loop transfer function $L(s)$ and associated sensitive's $P_s(s)$ and $P_t(s)$ are as follows:

$$L(s) = C(s)P(s), \tag{4.53}$$

where

$$P_s(s) = (I + L(s))^{-1}, \tag{4.54}$$

and

$$P_t(s) = L(s)(I + L(s))^{-1}. \tag{4.55}$$

---

41 The H_∞ methods described are based on applications as in [7].

42 More detail is available in [31].

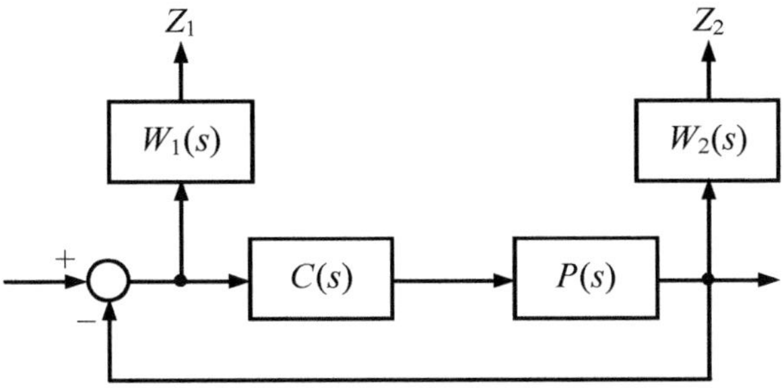

Fig. 4.19 System configuration with weights

The plant transfer function using state-space representation is

$$P(s) = C(sI - A)^{-1}B, \tag{4.56}$$

with

$$P_s(s) + P_t(s) = I. \tag{4.57}$$

The objective is to determine weights $W_1(s)$ and $W_2(s)$ so that the resulting controller is stable and

$$\left\| \begin{matrix} W_1(s)P_s(s) \\ W_2(s)P_t(s) \end{matrix} \right\|_\infty < 1. \tag{4.58}$$

Using the mixed sensitivity function,[43] which also includes solving two algebraic Riccati equations, the weights are determined[44] and a controller $C(s)$ is established. However, as noted earlier, the mixed sensitivity method proved to be less desirable, so that an alternate method using state feedback was considered.

The H_∞ algebraic Riccati equation method uses a state-feedback technique which is a lower order controller and which, unlike the mixed sensitivity approach, is less complicated and does not require cancelation of poles and zeros. The methods described in [32] are applied to solve an algebraic Riccati equation, from which the gains k and c (Fig. 4.5) are determined.

Smith Predictor Characteristics

The Smith predictor has been used with reasonable success to handle transport delays in various control applications [34]. A simple example to depict its use is in Fig. 4.20a,

[43] More detail of mixed sensitivity techniques is given in [33] and the references cited therein.

[44] For example, weight $W_1(s)$ is determined to make $P_s(s)$ small at lower frequencies, and $W_2(s)$ is determined to make $P_t(s)$ small at higher frequencies.

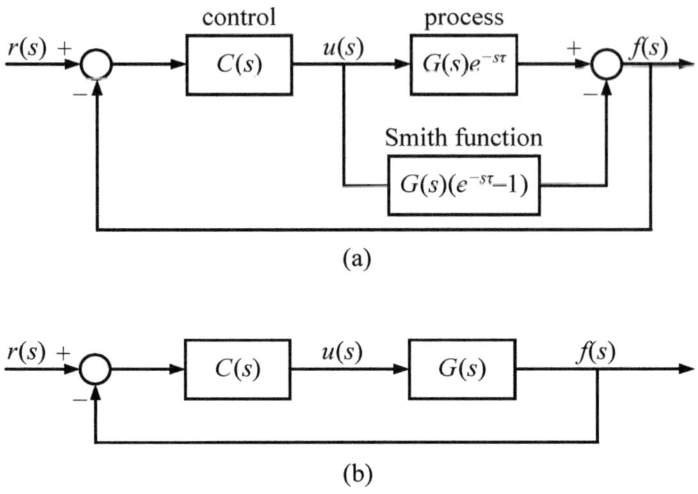

Fig. 4.20 **a** Application of Smith predictor, ideal case. **b** Application of Smith predictor, resulting model, ideal case

b. The basic idea is to develop a model that cancels the transport delay (τ), but retains the initial control. This is depicted in Fig. 4.20b, wherein the feedback $f(s)$ to the control $C(s)$ is not delayed by the process, but ideally the control output occurs downstream in the process after the delay.

What is depicted as the Smith predictor is a basis for the specific uses as noted previously for various applications herein.

Derivation of Mass Flow Techniques

The applicable portions of the model that supports the description of the mass flow techniques are as follows.

From Chap. 2 (2.1), the specific roll force using Sims' model is

$$P = \left(kQ_{\mathrm{p}} - \overline{\sigma}\right)\sqrt{R_{\mathrm{p}}\delta}, \tag{4.59}$$

where k is the is the resistance to deformation (hardness) of the strip, and δ is the stand draft, i.e., $\delta = h_{\mathrm{in}} - h_{\mathrm{out}}$, where h_{in} and h_{out} are the input and output strip thicknesses of the stand; the other elements of (4.59) are as in (2.1) and its supporting descriptions in Chap. 2.

In the linearization process, for smaller excursions around the operating point, the change in the specific roll force is approximated as

$$\Delta P = \alpha_1 \Delta k + \alpha_2 \Delta Q_{\mathrm{P}} + \alpha_3 \Delta\overline{\sigma} + \alpha_4 \Delta R_{\mathrm{P}} + \alpha_5 \Delta\delta. \tag{4.60}$$

For the purposes of this work, the following terms are applicable, where

$$\alpha_1 = \frac{\partial P}{\partial k}, \quad \alpha_6 = \frac{\partial P}{\partial h_{\text{in}}}, \quad \alpha_7 = \frac{\partial P}{\partial h_{\text{out}}}, \tag{4.61}$$

and

$$\Delta P_{\text{m}} = \alpha_1 \Delta k + \alpha_6 \Delta h_{\text{in}} + \alpha_7 \Delta h_{\text{out}}. \tag{4.62}$$

From (4.16),

$$h_{\text{out}} = \frac{F_{\text{m}}}{M} - S_{\text{m}}, \tag{4.63}$$

and (4.18) for stand $i + 1$

$$h_{\text{in},i+1} v_{\text{in},i+1} = h_{\text{out},i+1} v_{\text{out},i+1}. \tag{4.64}$$

Combining above the sensitivities of the entry speed of the trim at the upstream stand are:

$$\Delta v_{\text{in},i} = \lambda_1 \Delta h_{\text{in},i+1} + \lambda_2 \Delta S_{\text{m},i+1} + \lambda_3 \Delta k_{i+1}, \tag{4.65}$$

where λ_1, λ_2, λ_3, are determined similarly as in [15] and the references cited therein, for the sensitivity for each strip being processed.

Brief Overview of Kalman Filter as a State Estimator

As depicted in Fig. 4.12, the Kalman filter is used for the estimation of states. What follows is a brief overview to provide some basic insight into the filter's characteristics as a state estimator which relates well to this application. As noted previously, the overall function of the Kalman filter is to assure usable transmissions of data plus accurate tracking between various variables. The Kalman filter is very well-known and has had considerable successful usage in a great variety of applications over many years.

Assuming a state-space model of a linear system

$$\frac{\mathrm{d}x}{\mathrm{d}t} \equiv \dot{x} = Ax + Bu + w_{\text{d}}, \quad x(0) = x_0, \tag{4.66}$$

$$y = Cx + w_{\text{n}}, \tag{4.67}$$

where (A, B) is controllable, (A, C) is observable, w_d represents disturbances, is Gaussian, with covariance V_d, and w_n represents noise, is Gaussian, with covariance V_n.

The Kalman filter is based on the model (4.66) and (4.67); with the objective of determining the estimated states x_{est}, using u and y as inputs. The basis of the estimation is the relationship of the matrices V_d and V_n as used in an algebraic Riccati equation (ARE).

The error in estimation is determined by evaluating the equation:

$$\dot{\varepsilon} = (A - K_f C)\varepsilon, \quad \varepsilon(0) = \varepsilon_0, \tag{4.68}$$

where

$$\varepsilon = (x - x_{est}), \tag{4.69}$$

K_f is the Kalman gain, and where K_f is determined as the solution to the ARE to minimize a cost function J, and where

$$J = E\left[(x - x_{est})'(x - x_{est})\right], \tag{4.70}$$

E is the expected value, (′) is the matrix transpose, and where V_d and V_n are weighting matrices in the ARE. Thus the expected error between the states and the estimated states is minimized.[45]

Additional Detail of the Multivariable Control

The inputs and outputs[46] of the Kalman filter K_f and the gain matrix K_g are as shown in Fig. 4.21.

---

[45] More detail is available in [15, 18].

[46] Main signals are depicted; additional supporting signals are described in [18].

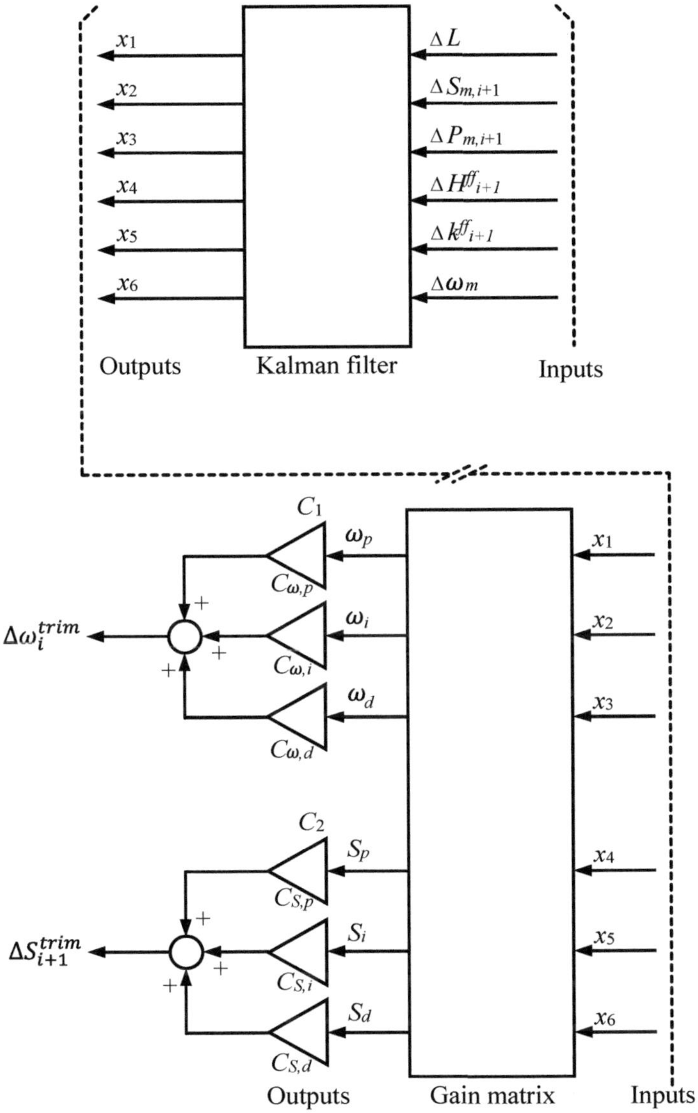

Fig. 4.21 Kalman filter with inputs and outputs (Tables 4.3 and 4.4). Gain matrix with inputs and outputs[47] (Tables 4.4 and 4.5)

Example of Eccentricity[48]

See Figs. 4.22 and 4.23

---

[47] Output trims $\Delta\omega_i$ and ΔS_{i+1} are denoted as u_r and u_c in Fig. 4.12.

[48] This example is also depicted in Chap. 3, Sect. 3.4.2.

Table 4.3 Kalman filter inputs

Input[49]	Description
ΔL	Strip length between stands i and $i+1$, using measure of looper angle and looper geometry
$\Delta S_{m,i+1}$	Measured hydraulic cylinder position, stand $i+1$
$\Delta P_{m,i+1}$	Measured roll force, stand $i+1$
ΔH_{i+1}^{ff}	Feedforward estimate, entry strip thickness stand $i+1$
Δk_{i+1}^{ff}	Feedforward estimate, entry strip hardness, stand $i+1$
$\Delta \omega_m$	Measured roll rotational speed difference, relative to stand $i+1$, based on stand set-up speeds

Table 4.4 Kalman filter outputs/gain matrix inputs

Output/ input[50]	Description
x_1	Strip length between stands i and $i+1$, using looper angle and looper geometry
x_2	Hydraulic cylinder position, stand $i+1$
x_3	Roll rotational speed, stand i
x_4	Entry strip thickness disturbance, stand $i+1$
x_5	Entry strip hardness disturbance, stand $i+1$
x_6	Strip speed disturbance

Table 4.5 Gain matrix outputs

Output	Description
ω_p	Stand i roll rotational speed trim, proportional reference ($C_{\omega,p}$)
ω_i	Stand i roll rotational speed trim, integral reference ($C_{\omega,i}$)
ω_d	Stand i roll rotational speed trim, derivative reference ($C_{\omega,d}$)
S_p	Stand $i+1$ cylinder position trim, proportional reference ($C_{S,p}$)
S_i	Stand $i+1$ cylinder position trim, integral reference ($C_{S,i}$)
S_d	Stand $i+1$ cylinder position trim, derivative reference ($C_{S,d}$)

[49] The notation Δ implies a deviation from set-up values.

[50] Estimated system states using Kalman filter as an observer. The notation Δ is applicable for each state.

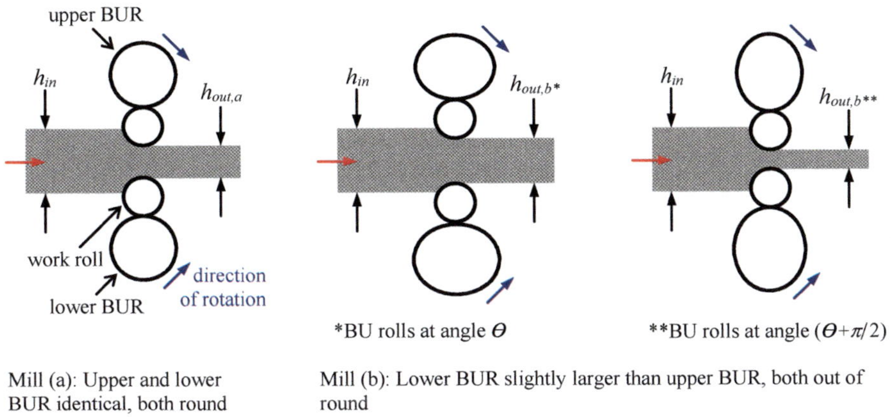

Fig. 4.22 Mill **a** without eccentricity; and mill **b** with eccentricity, where $h_{\text{out,b}}$ changes as BU rolls rotate from angle Θ to $(\Theta + \pi/2)$

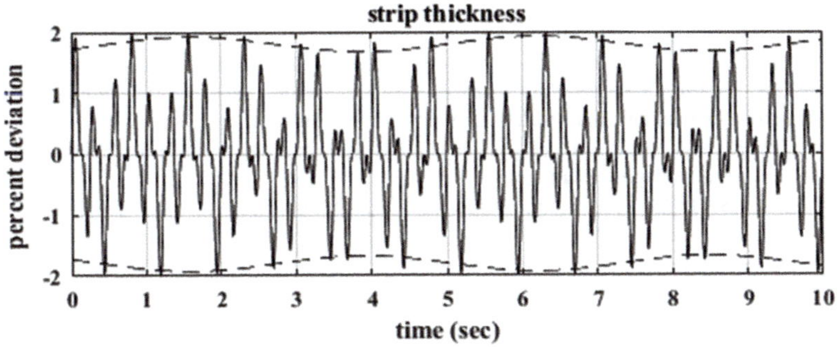

Fig. 4.23 Typical effect of BUR eccentricity on the strip thickness, also depicting a beat frequency effect (dashed lines)[51]

References

1. Choi et al (2007) Looper and tension control in hot rolling mills. J Process Control 17:509
2. Price JC (1973) The hot strip looper system. IEEE Trans Ind Appl 9(5):551–562
3. Rosenbrock HH (1974) Computer aided control system design. Academic Press, London
4. Kotera Y, Watanabe F (1981) Multivariable control of hot strip mill looper. In: Proceedings of the 1981 eighth world congress XVIII, pp 1–6
5. Hamada K et al (1985) Finishing mill tension control system in the Mizushima hot strip mill. Kawasaki Steel Tech Rep 11:35–43
6. Seki Y, Sekiguchi K et al (1991) Optimal multivariable looper control for hot strip finishing mill. IEEE Trans Ind Appl 27(1):124–130
7. Imanari et al (1997) Looper H-infinity control for hot strip mills. IEEE Trans Ind Appl 33(3):790–796

[51] Adapted from [23]. Copyright 1993, Marcel-Dekker, courtesy Taylor-Francis (all rights reserved). Figure developed using MATLAB®.

8. Ferguson IK, De Tina RR (1986) Modern hot-strip thickness control. IEEE Trans Ind Appl IA-22(5):934–940
9. Egawa N et al (1999) Hot rolling technology for producing high quality stainless steel at No. 3 hot strip mill in Chiba works. Kawasaki Steel Tech Rep 40:16–22
10. Shun-ichi N et al (2005) Technology and products of JFE steel's three plate mills. JFE Tech Rep 5:1–9
11. Bilkhu TS (2010) Hot strip mill gauge control. In: Proceedings of UKACC international conference on control, Coventry, pp 150–155
12. Kugi A, Schlacher K et al (2001) Nonlinear control in rolling mills: a new perspective. IEEE Trans Ind Appl 37(5):1394–1402
13. British Iron and Steel Research Association. UK patent 713–105
14. Prinz K et al (2019) Online parameter estimation for adaptive feedforward control of the thickness in a hot strip rolling mill. J Manuf Sci Eng 141(7):071005-1–071005-12
15. Hearns G, Bilkhu TS (2012) Mass flow control for intercritical rolling in hot mills. In: GE energy, power conversion technical article, Rugby, UK
16. Stephen RI, Randall A (1997) On-line adaptive control in the hot rolling of steel. IEE Proc Control Theory Appl 144(1):15–24
17. Sorenson HW et al (1985) Kalman filtering: theory and applications. IEEE Press, New York
18. Hearns G et al (2004) Hot strip mill multivariable mass flow control. IEE Proc Control Theory Appl 151(4):386–394
19. Pittner J, Simaan MA (2011) Tandem cold metal rolling mill control, chap 5.2. Springer, London
20. Hearns G et al (2005) Application of multivariable and predictive control techniques in modern hot strip mills. In: Iron and steel technology conference proceedings, Jan 2005, pp 59–67
21. Fryer CA, Hearns G (2009) Hot strip mill models and control for ¼ standard gauge. In: Association for iron and steel technology conference proceedings, vol II, pp 99–105
22. Teoh E et al (1984) An improved thickness controller for a rolling mill. In: Proceedings of 9th IFAC congress, Budapest, Hungary
23. Ginzburg VB (1993) High quality steel rolling, chaps 11.1, 11.7, 12.11–12, 23.17, 27.19. Marcel Dekker, New York
24. Zipf M (2009) Innovations in shape measurement and control for cold rolled flat strip products. In: Ginzburg VB (ed) Flat-rolled steel products. CRC Press, Boca Raton
25. Auzinger D et al (2003) Application of advanced technology packages for improved strip profile and flatness in hot-strip-mills. Rev Metall 100(10):977–986
26. Guo RM (1996) Optimal profile and shape control of flat sheet metal using multiple control devices. IEEE Trans Ind Appl 32(2):449–457
27. Guo RM (1998) Development of an optimal crown/shape level-2 control model for rolling mills with multiple control devices. IEEE Trans Control Syst Technol 6(2):172–179
28. Domanti S et al (2009) Profile and flatness—non-interactive control concepts and design. In: Association for iron and steel technology conference proceedings, vol II, pp 957–968
29. Nan R et al (2005) Regulating crown and flatness during hot rolling: a multi-objective study using genetic algorithms. Metals Manuf Proc 20:459–478
30. Pittner J, Simaan MA (2011) Tandem cold metal rolling mill control, chap 4. Springer, London
31. Zhou K, Doyle JC (1998) Essentials of robust control, chap 4. Prentice-Hall, Upper Saddle River
32. Zhou K, Khargonekar PP (1988) An algebraic Riccati approach to H_∞ optimization. Syst Control Lett 11(2):85–89
33. Kwakernaak H (2002) Mixed sensitivity design. In: Proceedings of 15th IFAC triennial world congress, Barcelona, pp 61–66
34. Goodwin GC et al (2001) Control system design, chaps 7.4, 15.5. Prentice-Hall, Upper Saddle River

Chapter 5
Advanced Control

5.1 The Need for Advanced Control

The tandem rolling of hot metal strip is a nonlinear process that is highly complex, and with characteristics that change considerably during normal and abnormal operations. In addition, the final product is flat sheets of metal that have stringent requirements to conform to a desired geometry. As might be expected, this presents a significant challenge to the control engineer.

The conventional methods[1] of control generally have been quite successful in producing a product of reasonably good quality. Most of the conventional configurations are desirable to the users as they are based on single-input–single-output (SISO) proportional–integral (PI) linear strategies. Thus, these controllers are relatively easy to design and tune, since the design and commissioning personnel usually have limited backgrounds in advanced control theory. Additionally, since the control for processing material of variable characteristics has been fairly easy to design and adjust, design and commissioning costs were kept at relatively low levels.

However, while the SISO type control is generally easy to use, it has limitations that can preclude it from the performance necessary to satisfy the increasing demands of the market for tighter tolerances and more stringent requirements in other areas. Such limitations include the inability to deal fully with the dynamic interactions of various variables, the need to improve robustness to nonlinearities and parameter uncertainties, to better handle various disturbances, and others that are peculiar to certain processes. Thus the need arises for alternatives for improvement such as multi-input–multi-output (MIMO)-based systems. Such systems would retain the desirable characteristics of the SISO-based systems, but with the significant improvements needed to realize the performance required for conforming to the increasing demands for higher quality products, at reduced costs, and to others as required in particular situations.

[1] What is considered as a conventional method is as noted in Chap. 4.

J. Pittner and M. A. Simaan, *Tandem Hot Metal Rolling Mill Control*, Advances in Industrial Control, https://doi.org/10.1007/978-3-031-58082-6_5

In this chapter several advanced MIMO systems are considered as potential improvements over conventional methods for control of the tandem hot strip mill. It is felt that what is presented typically represents well-known methods wherein performance is established by thorough simulation or actual application.

5.2 SDRE Technique

The algebraic state-dependent Riccati equation (SDRE) technique is a practical and user-friendly method for control of larger nonlinear systems. This method was developed much earlier in the 1960s, and began to be developed only much later in the 1990s as being very useful. This occurred mostly as a result of major advancements in computational machines, which made the pointwise on-line solution of the algebraic Riccati equation feasible for processes that require fast updates of controller scans. Since then, the SDRE method continued to be very successfully applied to various larger nonlinear processes (e.g., [1]), and thus with considerable possibility for the useful control of tandem hot metal rolling mills.

To better appreciate this technique, a basic understanding of the linear quadratic regulator (LQR) method is necessary, and therefore some of the major characteristics of LQR are presented in what follows. It should be noted that the mathematics of the presentation are basic with the higher and more rigorous mathematical approaches available in the cited references.

5.2.1 Linear Quadratic Regulator

The general idea of the several characteristics of linear quadratic regulation is to obtain a control law, which when used for the control of a dynamic system, will minimize a performance index that represents a cost function. The minimization of this index then results in a controller that is optimal in the sense that undesired deviations of certain variables from their desired values are inherently reduced.

Additionally, the cost function can be constructed to limit the capability of the control effort required to perform the control action, and thus minimize the overall cost. In the case of the LQR, for purposes of this consideration, the coefficient matrices of the linearized model and of the cost function are taken to be constants. Thus the determination of the cost function is based on a linearized model of the process, and the term linear is used in the description of the regulator. Also, usually the cost function is a quadratic function of the state variables, and a quadratic function of the control variables; thus the term quadratic is used in the regulator's description. There are other possible functions that could be used; however, this usage is usually desirable as it gives an optimal controller with linear feedback, and a closed loop linear system.

The elements of the weighting matrices can be determined to emphasize certain variables of the system to make them more dominant during the action of the controller. Thus, as an example, by assignment of these elements, more weighting can be realized to emphasize the effect of certain elements assigned to the control versus those assigned to the states; e.g., by assigning weightings of certain elements a faster response in certain states can be realized at the expense of a greater control effort.

To help better understand the SDRE by paralleling with LQR, it is assumed that the LQR function is closed loop and always operates for a long period of time, as opposed for some other applications of LQR wherein a shorter period of operation, or open loop control, is used. In this case a linearized state equation and a quadratic cost function are used as in (5.1) and (5.2):

$$\frac{\mathrm{d}x}{\mathrm{d}t} \equiv Ax + Bu, \quad x(0) = x_0, \tag{5.1}$$

$$J = \frac{1}{2}\int_0^\infty \left(x'Qx + u'Ru\right)\mathrm{d}t, \tag{5.2}$$

where $x \in R^n$ is the state vector, $u \in R^m$ is the control vector, $A \in R^{n\times n}$ and $B \in R^{n\times m}$ are the coefficient matrices, $Q \in R^{n\times n}$ and $R \in R^{m\times m}$ are the state and control weighting matrices, J is the performance index that is to be minimized with respect to u, and ($'$) represents the transpose of a matrix.

The control law (5.4) that minimizes J is determined by the algebraic Riccati Eq. (5.3):

$$A'K + KA - KBR^{-1}B'K + Q = 0, \tag{5.3}$$

where the constant matrix $K \in R^{n\times n}$ is the solution to (5.3), and the control law is

$$u = -R^{-1}B'Kx. \tag{5.4}$$

The state Eq. (5.1) is a constraint equation that is to be satisfied concurrently with (5.2). To assure a solution to (5.3) the pair (A, B) must be stabilizable. To assure that solution, K is unique and is positive semi-definite, there must exist a matrix C such that $Q = C'C$ and that (A, C) is detectable. And K is positive definite if (A, C) is observable. Also, Q is to be positive semi-definite and R positive definite; and if Q is selected such that $\left(A, \sqrt{Q}\right)$ is observable, then the closed system is guaranteed to be stable.[2]

A simple example follows that illustrates the application of the LQR. In this example the linearized system is described in (5.5), (5.6) and (5.7):

[2] More detail on LQR and the understanding of the control and applicable terms (e.g., positive-definite) is in [2, 3].

$$\dot{x} = Ax + Bu, \quad x(0) = \left[-1\ 5\right]', \tag{5.5}$$

$$y = Cx, \tag{5.6}$$

where

$$A = \begin{bmatrix} 0 & 1 \\ 0 & 1 \end{bmatrix}, \quad B = \left[0\ 1\right]', \quad C = \left[\sqrt{10}\ 0\right]. \tag{5.7}$$

The elements of Q are selected to emphasize x_1, and taking $R = I$, so that

$$Q = \begin{bmatrix} 10 & 0 \\ 0 & 0 \end{bmatrix}, \quad R = \begin{bmatrix} 1 & 0 \\ 0 & 1 \end{bmatrix}. \tag{5.8}$$

The solution (positive definite) of the associated algebraic Riccati equation is

$$K = \begin{bmatrix} 7.9527 & 3.1623 \\ 3.1623 & 2.5149 \end{bmatrix}. \tag{5.9}$$

The resulting control law, as seen at u (Fig. 5.1),[3] is:

$$u = -R^{-1}B'Kx, \quad \text{or } u = -\left[3.1623\ 2.5149\right]x. \tag{5.10}$$

The resulting optimal feedback system is $u = -R^{-1}B'Kx$, where

$$\dot{x} = \hat{A}x, \quad y = Cx, \tag{5.11}$$

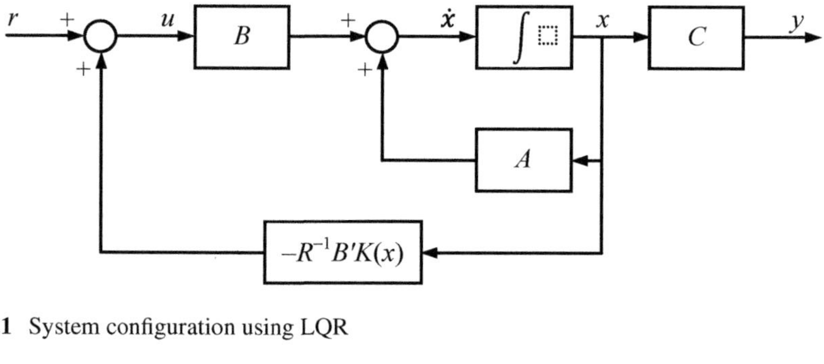

Fig. 5.1 System configuration using LQR

[3] As shown, the solution to the Riccati equation (i.e., the Kalman gain, K) is positive. The minus sign is shown as negative in the feedback box to be consistent with a positive sign for feedback in the summing junction with the reference. Often, the reverse is used; the sign in the feedback box is positive, and the sign in the summing junction is negative to be consistent with general control applications.

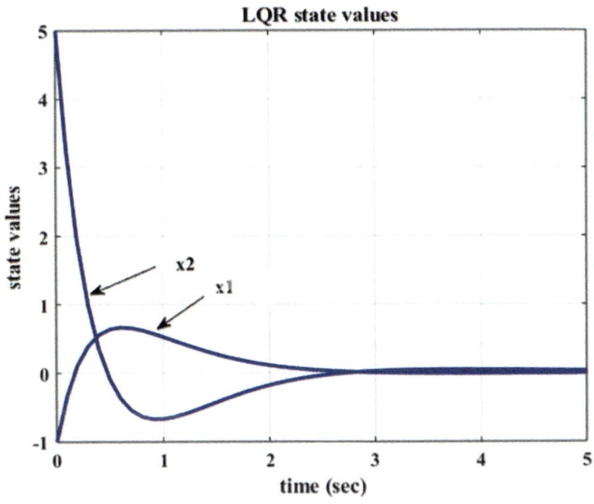

Fig. 5.2 State responses

$$\hat{A} = A - BR^{-1}B'K, \quad \text{or } \hat{A} = \begin{bmatrix} 0 & 1 \\ -3.1623 & -1.5149 \end{bmatrix}, \tag{5.12}$$

and r (= 0) is the system reference. The eigenvalues of $\hat{A}$ are $-1.2574 \pm j1.2574$, which implies closed loop stability of the system as shown in Fig. 5.1 which depicts the optimal feedback, and Fig. 5.2 which depicts the responses.

As noted previously, the controller acts to move the states from their initial values at $x(0)$ to zero, with a zero reference r. If the reference was not zero, appropriate adjustments would need to be made in the x and u terms in the cost function to account for the change in the operating point. This is shown later in this chapter in the uses of the SDRE method.

5.2.2 *SDRE Characteristics*

The basic structure of the state-dependent Riccati equation (SDRE) control is the same as the LQR control, with the exception that the matrices A, B, C, Q, and R are state-dependent. Since the basic structure of the SDRE is the same as that of the LQR controller, it might seem initially that this, from a theoretical standpoint, is a simple difference. However, this certainly is not so, as can be seen by the work as described in many recent publications, e.g., [4, 5]. Nonetheless, the usefulness of the SDRE method is quite well-known and has been highly successful in a variety of nonlinear applications [1]. This is mostly because it has many desirable characteristics that are unavailable in other nonlinear methods. The following brief overview is included to attempt to give some insight into the basic theory. More details and some recent advancements are available in the cited references.

It is assumed in a basic theoretical treatment that a nonlinear system can be described by a state Eq. (5.13) and an output Eq. (5.14):

$$\frac{dx}{dt} \equiv \dot{x} = a(x) + b(x)u, \tag{5.13}$$

$$y = g(x), \tag{5.14}$$

where $x \in R^n$ is a state-dependent vector that represents the individual state variables, $a(x) \in R^n$ is a state-dependent vector, $u \in R^m$ is a vector whose elements represent the individual control variables, $y \in R^p$ is a vector whose elements represent the individual output variables, $b(x) \in R^{n \times m}$ is a state-dependent matrix, and $g(x) \in R^p$ is a state-dependent vector. By factorizing $a(x)$ into $A(x)x$, $g(x)$ into $C(x)x$, and with $b(x) = B$, this becomes a form similar to linear state equations[4]:

$$\frac{dx}{dt} = A(x)x + Bu, \quad x(0) = x_0, \tag{5.15}$$

and

$$y = C(x)x. \tag{5.16}$$

Similar to the LQR problem, the objective in the SDRE method is to minimize the performance index J with respect to u as in (5.17), where

$$J = \frac{1}{2}\int_0^{\infty} \left(x'Q(x)x + u'R(x)u\right)dt, \tag{5.17}$$

subject to the constraint (5.15), where $Q(x)$ and $R(x)$ are weighting matrices, with $Q(x)$ semi-positive definite, $R(x)$ positive definite, and $a(x)$, $Q(x)$, $R(x)$ have continuous partial derivatives through the order k (≥ 1), for all x in the control space.[5] The intent of J is to determine a control law that will regulate the system to the origin.[6]

The solution is determined by finding a factorization of $a(x)$ such that (5.13) can be expressed in the form of (5.15). Then, on a pointwise basis, the state-dependent algebraic Riccati Eq. (5.18) can be solved to determine a solution $K(x)$:

$$A'(x)K(x) + K(x)A(x) - K(x)BR^{-1}B'K(x) + Q(x) = 0. \tag{5.18}$$

[4] The theoretical requirements for factorization as described in [6] are met. Thus $A(x)$ (always existing and non-unique), and $C(x)$ are state-dependent matrices, and for this application B is a constant (non-zero) matrix.

[5] Herein "all x" is understood to mean "all x in the control space."

[6] As noted previously in the description of LQR, wherein there is a non-zero reference, the objective is to regulate to the desired reference.

The resulting solution is used in the control law:

$$u = -R^{-1}(x)B'K(x)x, \tag{5.19}$$

where $K'(x) = K(x)$ for all x, and $K(x)$ is positive semi-definite. In addition, for stability, the pair ($A(x)$, B) should be pointwise stable for all x.

In general, except for certain special cases, asymptotic stability for all x has to be verified by simulation, as there is no useful theory that covers mostly all situations. However, local asymptotic stability can be realized in certain special applications as in [7]. Also, the necessary conditions for optimality in certain special cases are as addressed in Section "Optimality of Certain Applications" of Appendix. In general these conditions are satisfied only if $A(x)$ is unique, which is usually unknown at the start. However if $A(x)$, $K(x)$, $Q(x)$, $R(x)$ and their gradients with respect to x are bounded for all x, and if there is global asymptotic stability, the states will converge to optimal states as they are driven to their final state.

It also can be noted that, although the eigenvalues for all x in a certain closed loop system as controlled by the SDRE technique are in the open left half plane, global asymptotic stability may not exist. An interesting example is presented in Section "Stability in Open Left Half Plane" of Appendix.

Thus, generally overall there are few theoretical bases; and yet there are concerns regarding global asymptotic stability, prediction of performance, robustness to disturbances and uncertainties, optimality, and others. Thus in many instances such concerns must be addressed and resolved by simulation.

Moreover, in the instance of the tandem hot strip finishing mill, the process is large, there are significant nonlinearities, large time delays, major uncertainties and disturbances, and others as noted earlier in Chap. 3. Additionally, there is the requirement for the controller to be user friendly to design, commissioning, and maintenance personnel that have limited backgrounds in advanced control theory; plus the need for improvement in quality due to tighter requirements in customer specifications, rapid start-up to increase production, and the reduction of costs. What follows shows how these issues, and others, are addressed and resolved in the augmented SDRE technique.

5.2.2.1 Basic Controller Structure

As noted previously, the control of the tandem hot metal rolling process is especially challenging. To provide some additional insight into the SDRE method, it is felt that some presentation of the basic control structure and its functioning might help to better understand the present characteristics of the controller and its advantages.

In the case of the SDRE, one of the main functions of the controller is to reduce excursions in the centerline thickness and tension in the strip from their desired values. The realization of this requires the operating point to be shifted to a non-zero value, which corresponds to the desired operating point of the process, and which is done by changing the coordinates as in (5.20) and (5.21):

$$z = x - x_{\text{op}}, \tag{5.20}$$

where $z \in R^n$, the state vector is $x \in R^n$, and $x_{\text{op}} \in R^n$ is a vector whose elements represent the individual states at the operating point. The cost function then becomes

$$J = \frac{1}{2}\int_0^\infty \left(z'Qz + \left(u - u_{\text{op}}\right)'R\left(u - u_{\text{op}}\right)\right)\mathrm{d}t, \tag{5.21}$$

where $u \in R^m$ is the control vector, $u_{\text{op}} \in R^m$ is a vector whose elements represent the individual control variables at the operating point, and $u_{\text{f}} \in R^m$ (Fig. 5.3) is the feedback vector. During operation some of the elements of x_{o} and u_{op} will change depending on the operation. Additionally, the weighting matrices Q and R are taken as diagonal with constant elements that are tunable and may be changed during different operations and depending on the particular situation.

Moreover, the controller must significantly reduce the effects of internal and external disturbances, and do so in the presence of uncertainties in modeling and measurements. The more significant of these are described in Chap. 3. In the case of SDRE, both the disturbances and uncertainties are represented by changes in the elements of the $A(x)$ and $C(x)$ matrices, so that (5.15) and (5.16) become

$$\dot{x} = A(x)x + \delta A(x)x + Bu, \quad x(0) = x_0, \tag{5.22}$$

and

$$y = C(x)x + \delta C(x)x, \tag{5.23}$$

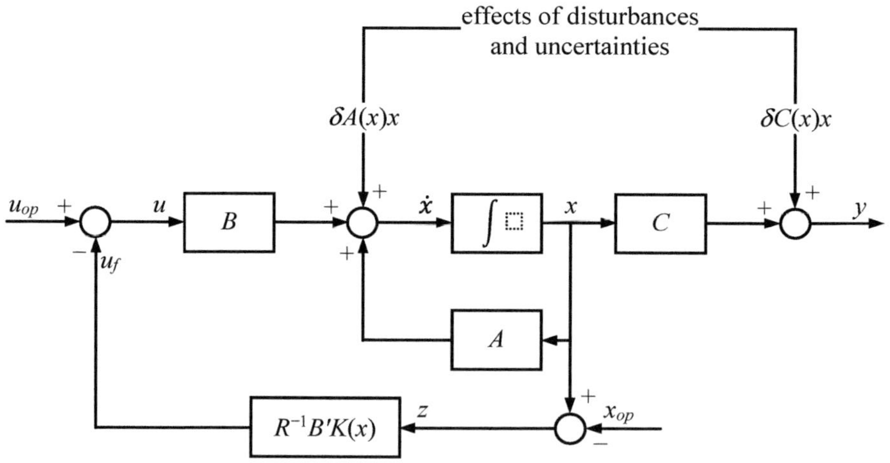

Fig. 5.3 Control without augmentation

where $\delta A(x)$ and $\delta C(x)$ are matrices whose elements are the deviations in the elements of $A(x)$ and $C(x)$. An initial system is depicted in Fig. 5.3.

During initial work it was determined that some augmentations to the initial system were needed to realize a more useful result. This was based on simulations using data from actual applications, and wherein several possible methods were considered.

Among the difficulties were the handling of disturbances and uncertainties, e.g., disturbances in incoming thickness and hardness, in modeling uncertainties, and in methods for simulating interstand time delays, plus others. Certain methods required a considerable number of additional states, and without significant improvements. A much simpler method was the addition of outer loops for tension and thickness which resulted in considerable improvements, such as significant reduction in the effects of disturbances and uncertainties, reasonable ease of implementation, and good performance during various modes of operation.[7]

5.2.2.2 Initial Improvements

A system with outer loops as an augmentation (Fig. 5.4) was evaluated, with reasonably good success. The augmentation included trims for strip thickness and tension. A PI control with proportional and integral gains K_I and K_P was added for the thickness trims, as was a suitable function φ_r for the trims in tension, and φ_y for the trims in thickness.

This configuration was initially intended for use with the tandem cold rolling process; however, with slight variations, it also has characteristics that are useful as a basis for tandem hot rolling, which is described further in Sect. 5.2.3.

5.2.3 *Application to Tandem Hot Rolling*

The application of SDRE to the tandem hot strip mill is similar to what was developed for the tandem cold strip mill, as described in the aforementioned. Exceptions are revisions and additions as needed to handle the control of the loopers, the differences required for processing material of a higher temperature, the lack of measurement of significant variables due to a hostile environment, plus others resulting from the differences in processing. An initial controller is described in what follows in Fig. 5.5a, b.

The overall control is depicted in Fig. 5.5a, with tension and thickness trims, and with algorithms φ_r for looper position and φ_y for strip thickness and interstand tension. In Fig. 5.5a φ_y is an algorithm that generates y_c, which is a vector whose elements are the elements of y, either measured or estimated. The K_P and K_I blocks are diagonal matrices whose elements are constants that are gains for the PI thickness and tension trims.

[7] More detail at this stage of development is available in [8].

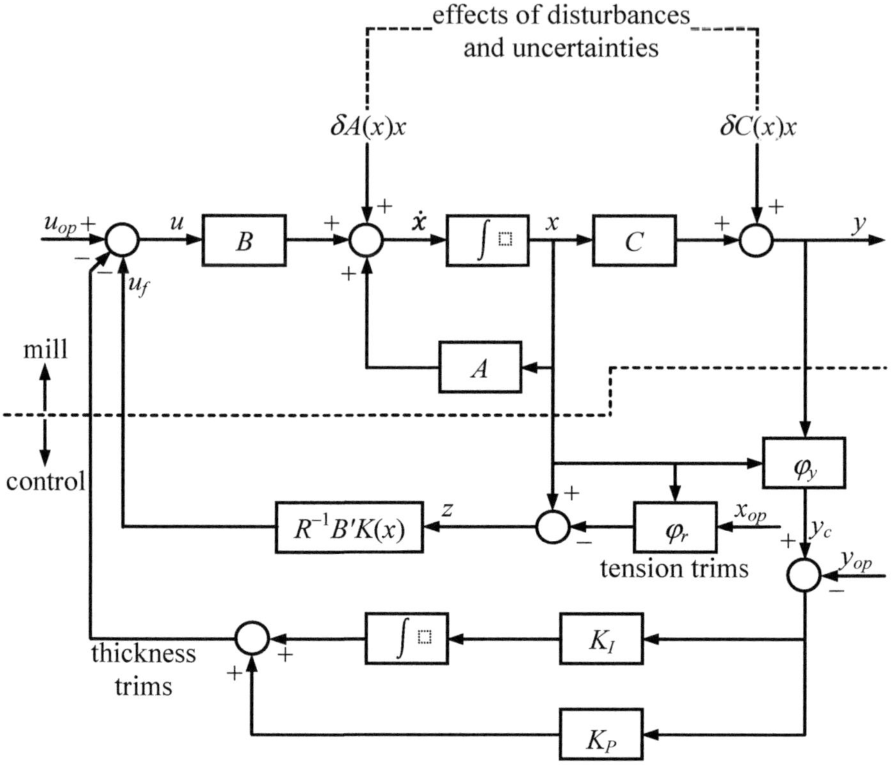

Fig. 5.4 Initial control with augmentation

Figure 5.5b represents an algorithm φ_r in Fig. 5.5a that generates the looper operating point trims. In Fig. 5.5b $x_{op,i}$ is an element of the vector x_{op} that represents the desired operating point of the looper i that is between stands i and $i + 1$, x_i is the measured position of looper i, and $K_{i,\,i+1}$ is a gain term for stands i and $i + 1$. A direct feed through is for the remaining elements of x_{op}.

In the control of the tandem hot mill, to realize a high quality of the final output and the stability of rolling, it is essential to reduce excursions in the looper positions, the interstand tensions, the strip thicknesses, and others. Additionally, this must occur within a hostile mill environment which precludes the valid measurement of significant variables. Some of these are the strip speeds, output thicknesses at the first and intermediate stands, strip temperatures, and other material properties of the strip, such as resistance to deformation (i.e., hardness).

To deal with this situation, the controller is configured so that the excursions in unmeasured variables can be reasonably determined by the use of measured variables, the models, and their refinements. These refinements generally use data collected during actual rolling, and other techniques; wherein, for example, beneficial adjustments are made to the actual measurements and models. Most of these are done in level 2 as part of data collection and the refinement of models and measurements for use by the controller in level 1. Additionally, the nonlinear MIMO nature of the

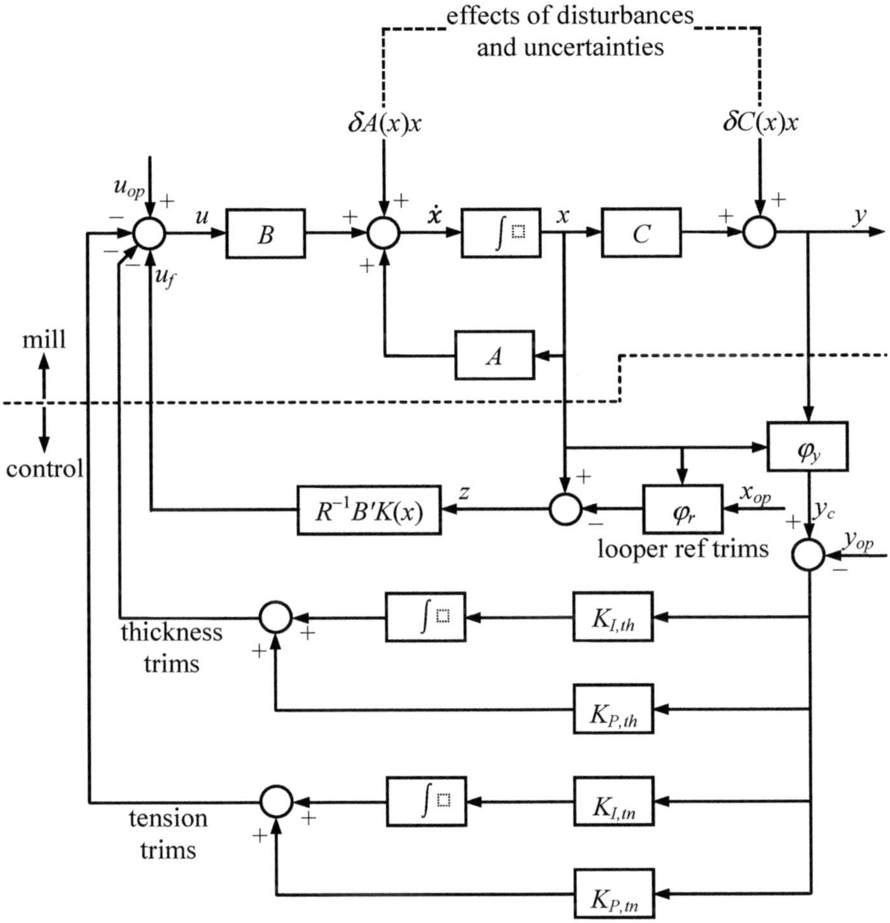

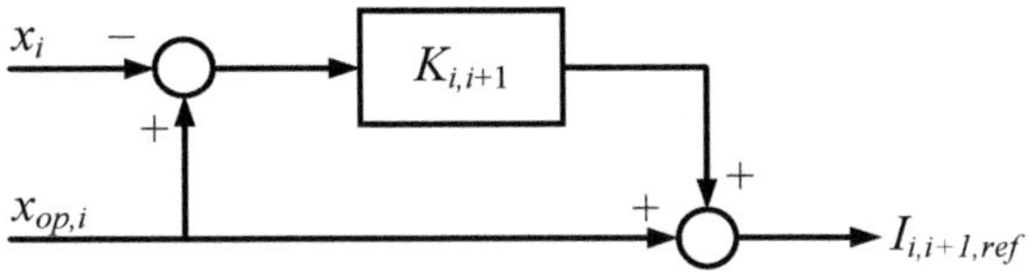

Fig. 5.5 **a** SDRE hot mill control with augmentation. **b** SDRE hot mill looper angle operating point trim[8]

---

[8] Copyright IEEE 2010, Fig. 5.5a, b as shown in this section. Portions of these are reprinted with cosmetic alterations, by permission, from [9].

Table 5.1 Mill operating point

Stand	h_{out} (mm)	T (°C)	V_0 (m/s)	R (mm)
Entry	38.8	1058	–	–
1	21.6	988	1.188	360
2	14.4	973	1.823	336
3	8.6	957	2.957	353
4	6.1	938	4.294	343
5	4.7	922	5.665	388
6	3.9	904	6.946	348
7	3.5	894	7.880	369

SDRE helps to inherently handle the interactions between variables, both measured and unmeasured. Further, the closed loop nature of the SISO trimming functions can reduce slight deviations in the looper positions, as well as providing zero steady-state error in measurements, reducing the effects of interstand time delays, and providing a user-friendly means of controller settings to facilitate start-up as needed. All of this significantly enhances the robustness of the controller to these effects.

At this stage, initial design and simulations were performed to verify the controller's performance. The data used were typical of actual usage as described in Table 5.1.[9]

As part of the initial design effort, the following were considered as possible uncertainties and disturbances:

Modeling Uncertainties

- mill entry thickness: ± 5%;
- strip temperatures (Table 5.1): ± 10 °C;
- strip carbon content: ± 30%;
- mill modulus: ± 10%.

Measurement uncertainties in strip tension stress[10] were taken to be about ± 1%. Others, taken as ± 0.1% or less, were roll gap cylinder position, work roll peripheral speed, roll force, and looper torque, angle, and angular velocity. During initial simulations these uncertainties were shown to be less significant. Additionally for this initial evaluation, uncertainties in strip temperatures and carbon content are assumed to be constant for the entire length of the strip, and Young's modulus, strip density, and width are taken to be constant.

The disturbances considered were external and internal. External disturbances included changes in incoming thickness and temperature due to skid chill in the reheating furnace. The effects of these disturbances were estimated as they were tracked through the mill and combined with data in Table 5.1:

[9] Data in Table 5.1 are from actual measurements as noted in Chap. 2. Other data as noted are based on manufacturer's information for typical mill operation.

[10] Tension stress measurement is based on measurement of looper parameters.

Typical Disturbances in Entry Thickness and Temperature

- mill entry thickness: + 5%;
- strip temperatures (Table 5.1): – 15 °C.

Additional assumed external disturbances were a fast increase of about 20% in the carbon content at the mill entry. Internal disturbances were undesirable perturbations in the strip speed and looper torque at each mill stand, in accordance with (5.24) and (5.25):

$$v(t) = V_0\Big(0.05 + 0.005\Big(\sin(4t) + \frac{\pi}{6}\Big) + \sin(7t)\Big), \tag{5.24}$$

and

$$m(t) = M_0\Big(0.05 + 0.005\Big(\sin(4t) + \frac{\pi}{6}\Big) + \sin(7t)\Big), \tag{5.25}$$

where V_0 (m/s) and M_0 (kN m) are respectively the operating points for the strip speed and looper torque for each stand. Further, it was assumed that compensation for roll eccentricity was quite effective so that its effect was less significant.[11] Also considered was a ramp increase in mill speed (i.e., mill zooming) of about 20% in 10 s above operating point speed, which is part of typical normal operations for certain products and typical mills a bit older. Simulations were performed to verify the previous design characteristics. The simulations were performed for a fully threaded mill, operating at a constant speed, and then with a ramped increase in the speed by about 20% above the constant speed to simulate mill zooming. The settings of the constant diagonal matrices Q and R, and the settings of the trim functions, were made intuitively which was confirmed by the simulation. The uncertainty in the mill modulus M was taken to be within about 2% based on active compensations available via level 2 activity.

Cases were simulated with various combinations of concurrent uncertainties in the mill modulus, interstand tension stresses and looper positions. Also included were uncertainties of 5% in the incoming thickness, in the carbon content of 30%, and in the temperature of 10 °C. The various disturbances and uncertainties were addressed individually, and then combined to approximate the more severe scenarios. Some typical results of simulations are as follows:

*Responses to Uncertainties with Disturbances***

- tension stress: 1.1%;
- looper position: 0.1%;
- exit thickness: 0.7%;
- exit thickness, stand 7: neg.

** Maximum magnitude of percent deviation from operating point values

[11] Compensation of roll eccentricity is addressed later in this chapter.

Additional verification was done to confirm the time required for the on-line solution of the algebraic Riccati equation (ARE). In this case a solution was determined using the matrix sign function [10], and also by using a corresponding MATLAB® function for solution of the ARE. Considering the similarity between the capabilities of the hardware and software used in the simulations to those in actual applications, it was determined that the solution of the ARE was within the needed scanning capability of the then typical standard available commercial systems, i.e., within 10–15 ms.[12]

5.2.4 Further Improvement in the SDRE Method

As noted previously the SDRE technique can offer a significant and useful improvement in the control of the tandem hot metal rolling process. The work described previously has been expanded to consider further advancements. These are the deletion of (1) the need for look up tables for basic functional characteristics of various system configurations, and (2) the necessity for solving on-line the state-dependent Riccati equation, and without degradation in performance. It is considered that this represents a significant advancement in the control of this process and has the possibility for improvement in other similar applications.

The improved controller has three regimes of operation, (1) the basic regime (BA), (2) the pre-roll regime (PR), and (3) the roll regime (r). In the basic regime a basic model of the mill is developed. The main idea is to keep the dynamics of the inner control loop unchanged between the basic loop and the pre-roll loop.

The basic regime is kept reasonably updated for the various types of product being processed, while the pre-roll regime is further updated based on the best known data of the upcoming piece of a particular product to be immediately processed. The control law gain K_{BA} of a particular product of the basic regime is determined by the algebraic Riccati equation (ARE) off-line at the state x_0:

$$K_{\mathrm{BA}}(x_0) = R^{-1}B'K(x_0), \tag{5.26}$$

where $K(x_0)$ is the solution of the ARE at x_0. Thus the controller of the inner loop plus the gains of the PI trim loops can be used as a base controller from which variations in the particular piece of a product can be derived using the pre-roll regime.

The settings of the controller just before the strip enters the first mill stand are established by the pre-roll regime. The separate model used in this regime is based on the most recent data during processing that reasonably describes the upcoming piece. This, combined with the basis established in the basic regime, is used to determine the gain of the inner loop of the controller and the settings of the PI outer loop trims. The intent is to keep the dynamics of the inner loop of the pre-roll regime the same as the basic regime without needing to solve the ARE. For the inner loop this is done

[12] In what follows it is shown that the on-line solution of the ARE would be unnecessary.

by matching the ordinary differential equations (ODEs) of the two loops, e.g., the ODEs of the pre-roll and the basic regimes. The closed loops are

$$\dot{x} = (A_{\mathrm{PR}}(x) - BK_{\mathrm{PR}}(x))x + Bw, \quad x(0) = x_0, \tag{5.27}$$

and

$$\dot{x} = (A_{\mathrm{BA}}(x) - BK_{\mathrm{BA}}(x))x + Bw, \quad x(0) = x_0. \tag{5.28}$$

The $A_{\mathrm{BA}}(x)$ and $K_{\mathrm{BA}}(x)$ correspond to $A_{\mathrm{PR}}(x)$ and $K_{\mathrm{PR}}(x)$, so that the dynamic characteristics of the these will be equivalent if

$$(A_{\mathrm{PR}}(x) - BK_{\mathrm{PR}}(x))x + Bw = (A_{\mathrm{BA}}(x) - BK_{\mathrm{BA}}(x))x + Bw, \quad x(0) = x(0), \tag{5.29}$$

and thus $K_{\mathrm{PR}}(x_0)$ is determined as

$$K_{\mathrm{PR}}(x_0) = B^{-L}(A_{\mathrm{PR}}(x_0) - A_{\mathrm{BA}}(x_0)) + K_{\mathrm{BA}}(x_0), \tag{5.30}$$

where $A_{\mathrm{PR}}(x_0)$ is determined from the updated model at the operating point of the pre-roll regime, $A_{\mathrm{BA}}(x_0)$ and $K_{\mathrm{BA}}(x_0)$ are as previously noted, and $B^{-\mathrm{L}}$ is the left inverse[13] of the B matrix of the mill model. Thus the dynamics of the inner control loop are essentially unchanged from those of the basic regime. And also the dynamics of the PI trims of the pre-roll regime are set to keep the overall characteristics of both of the outer loops to be very similar to each other. Additionally, $K_{\mathrm{PR}}(x_0)$ is the starting point of the initial scan of the roll regime.

As the strip is processed through the mill, the objective of the roll regime is to adjust the controller to maintain the dynamic characteristics of the inner control loop, and those of the overall outer control loop, close to the characteristics of the pre-roll dynamics. As in many applications of the SDRE, the setting of the inner control loop feedback gain is determined pointwise during each controller scan.

For example, at the beginning of a scan i the variables represented by the state vector are measured (or otherwise evaluated). The value of the $A_{\mathrm{r},i}(x)$ matrix is computed later in the scan. By matching with the ODE of the previous scan $i - 1$, the dynamics of the present scan i will remain essentially unchanged from the dynamics of the previous scan. This will occur if the feedback gain $K_{\mathrm{R},i}(x)$ for this scan is then determined as in (5.31), wherein $A_{\mathrm{R},i-1}(x)$ and $K_{\mathrm{R},i-1}(x)$ are as determined in the previous scan, i.e.:

$$K_{\mathrm{R},i}(x) = B^{-L}\big(A_{\mathrm{R},i}(x) - A_{\mathrm{R},i-1}(x)\big) + K_{\mathrm{R},i-1}(x). \tag{5.31}$$

[13] $B^{-\mathrm{L}}$ is the left inverse of the B matrix, which exists and is computed as $B^{-\mathrm{L}} = (B'B)^{-1}\, B'$.

In the first scan of the controller under the roll regime, the matrices $A_{R,i-1}(x)$ and $K_{R,i-1}(x)$ are replaced with $A_{PR}(x_0)$ and $K_{PR}(x_0)$, as determined in the pre-roll regime.

Additionally, to keep the characteristics of the outer loop nearly invariant, the settings of the gains of the PI trims are determined similarly using pertinent elements of the $C(x)$ matrix. However, this adjustment may be unnecessary in certain cases.

Simulations were done similar to those described in Sect. 5.2.3 with similar results. Additionally, the improvements noted can reduce the design effort since the pointwise solution of the ARE is not required for most applications. This reduces the scan time to be consistent with the requirements for viable process control. Also, any changes in the $A(x)$ and $C(x)$ matrices are measured (or otherwise determined), with automatic compensation to maintain the dynamic characteristics of the inner and outer control loops essentially invariant, so that less design and tuning efforts are required. Additionally the need for look up tables to modify characteristics of the controller to achieve desirable responses during rolling is reduced. Thus the overall results include improvements in rolling stability and product quality, with potential for significant reduction in costs.[14]

5.2.5 Response to Uncertainties and Disturbances[15]

The stability of rolling and the quality of the final product depends highly on the capability of the controller to successfully handle significant disturbances and uncertainties. This is especially so in cases of possible undesirable deviations in looper positions, interstand tensions, and strip thicknesses, plus others. In many cases these deviations are the results of the characteristics of incoming workpieces, uncertainties in modeling and measurements, and interactions between the large ranges of variables in the entire mill. The SDRE controller as described previously has been investigated to determine its capability to respond to these conditions. What follows describes the results of these initial efforts.

The controller is described in Sects. 5.2.3 and 5.2.4, wherein the main idea is to maintain the dynamic characteristics of the inner control loop, and the overall outer control loop, essentially invariant during the processing of the strip. Thus, the effects of undesirable conditions, such as disturbances and uncertainties, are considerably reduced. The structure of the controller is depicted in Fig. 5.5a, b, except with the SDRE feedback replaced by $K_R(x)$ as in (5.31). The $A(x)$ and $C(x)$ matrices are updated during every scan of the controller to realize automatic compensations as needed to handle disturbances and uncertainties to retain good performance.

---

[14] This method of control is patented [11].

[15] Copyright IEEE 2014, figures and tables as shown in this section. Portions of these are reprinted with cosmetic alterations, by permission, from [12]. Unless otherwise noted, data depicted are unchanged from [12].

Certain generally estimated uncertainties, without disturbances, in several variables are arranged in two groups as follows:

Group 1 variables		Group 2 variables	
Variable	Uncertainty	Variable	Uncertainty
α: tension stress	− 5%	α: tension stress	+ 5%
M: mill modulus	+ 10	M: mill modulus	− 10
h: strip thickness[a]	+ 10	h: strip thickness[a]	− 10
T: strip temperature[a]	− 20 °C	T: strip temperature[a]	+ 20 °C
θ: looper angle	+ 0.5	θ: looper angle	− 0.5
Carbon[a]	+ 30	Carbon[a]	− 30

[a] At the mill entry

The mill operating point is per Table 5.1 with an equivalent carbon of 0.3% (weight), without uncertainties. Disturbances, which can vary during the processing of the strip, are applied. Their resulting effects on the uncertainties are determined by simulation and are in addition to the listed uncertainties of Group 1 or Group 2. Additionally, it was assumed that the mill would be zoomed with an increase in mill speed of 20% over about 10 s starting at 3 s after the application of a disturbance. The results are listed as tables in [12]; Tables 5.2 and 5.3 as follows are examples.

Examples also are in Figs. 5.6 and 5.7, wherein there is a Group 1 and a Group 2 of uncertainty in the equivalent carbon of the strip at the mill input, followed by a disturbance of 40%. The effects of different strengths of the disturbance are also shown. The results are as summarized as a part of the entries in Tables 5.2 and 5.3

Table 5.2 Deviations in certain Group 1 variables[a]

Variables	Mean	Max	Standard deviation
$h_{out,1}$–$h_{out,6}$	− 1.9%	− 3.0	0.6
$h_{out,7}$	neg	neg	neg
σ_1–σ_6	5.0	5.0	0.0
θ_1–θ_6	− 0.1	− 0.1	0.0

[a] Deviations are from operating point values, under estimation of uncertainty and disturbances in equivalent carbon (Group 1)

Table 5.3 Deviations in certain Group 2 variables[a]

Variables	Mean	Max	Standard deviation
$h_{out,1}$–$h_{out,6}$	2.2%	3.3	0.6
$h_{out,7}$	neg	neg	neg
σ_1–σ_6	− 5.0	− 5.0	0.0
θ_1–θ_6	0.1	0.1	0.0

[a] Deviations are from operating point values, under estimation of uncertainty and disturbances in equivalent carbon (Group 2)

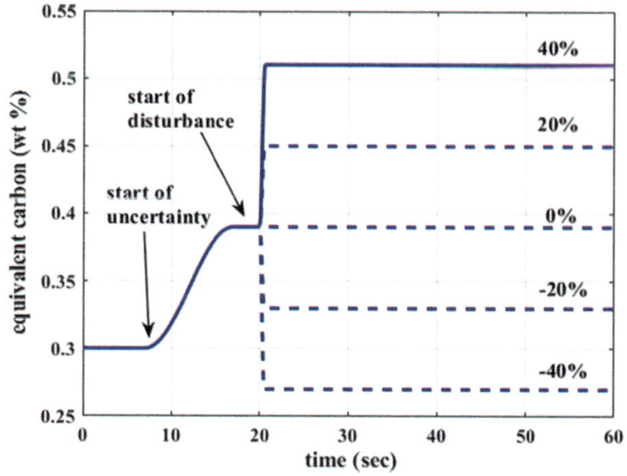

Fig. 5.6 Uncertainty and disturbance of equivalent carbon (Group 1)

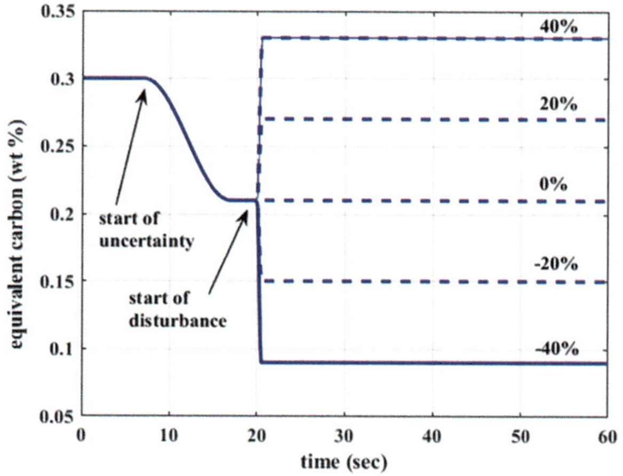

Fig. 5.7 Uncertainty and disturbance of equivalent carbon (Group 2)

and are intended to provide some insight into the capability of the initial control to reduce effects of various disturbances and uncertainties.

Another example is Fig. 5.8 which depicts the responses of the output thickness of stand 1 to various uncertainties and a disturbance in incoming thickness. This presents a basic example of the capability for rejection of disturbances in the presence of uncertainties.

The preceding shows a reasonably good capability for mitigating the effects of uncertainties and disturbances at a certain operating point. This, along with others in this section, presents a suitable basis for expansion of further improvements in this area.

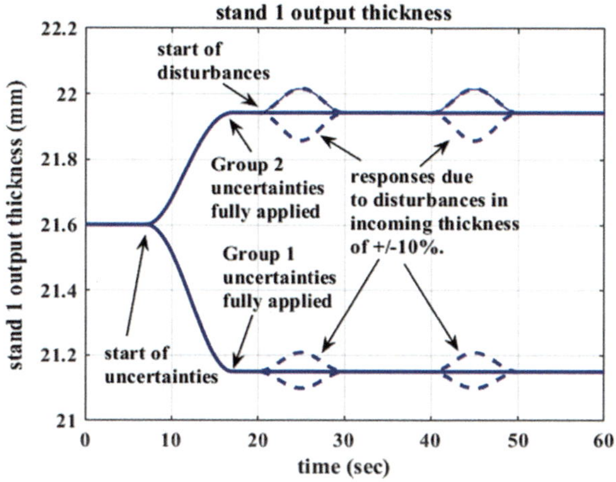

Fig. 5.8 Stand 1 output thickness with uncertainties and disturbances in incoming thickness

5.2.6 Reduction in the Likelihood of Cobbles[16]

In certain investigations, usually the material of the strip being processed is taken to be of a single phase,[17] such as austenite, and of medium and low carbon equivalents. It is well-known that during the processing of a strip of austenite, a significant decrease in the temperature of the strip could cause a change in the phase of the strip from austenite to ferrite. During this change in phase, the characteristic of the strip changes such that an increase to the resistance to deformation (i.e., the hardness) decreases as the temperature decreases, which is the opposite of what might be expected. This occurs during the change whenever there is a mix of the two materials and remains until the strip is fully ferrite.

This inversion of the strip characteristic has caused cobbles[18] in several instances, and especially wherein a decrease in temperature at a downstream stand is caused by a temperature disturbance at the mill entry as it would be moving through the stands. In many cases the disturbance results from skid chill in the annealing furnace and is sufficiently severe so that it is carried through the mill until the strip temperature decreases to where the change in phase occurs with the possibility of a cobble.

In certain conventional systems the inversion of the temperature characteristic is recognized by the controller as a decrease in the strip tension. This would cause the looper to increase its position to attempt to retain the desired tension, with an

[16] Copyright IEEE 2015, figures and tables as shown in this section. Portions of these are reprinted with cosmetic alterations, by permission, from [13]. Unless otherwise noted, data depicted are unchanged from [13].

[17] A "phase" of the strip material refers to its metallurgical characteristic; e.g., a change in phase could refer to a change from austenite to ferrite. During the change there is a mix of austenite and ferrite.

[18] A cobble is a wreck in the mill which results in a stop in production, the loss of the strip being processed, possible equipment damage, and could result in a costly recovery.

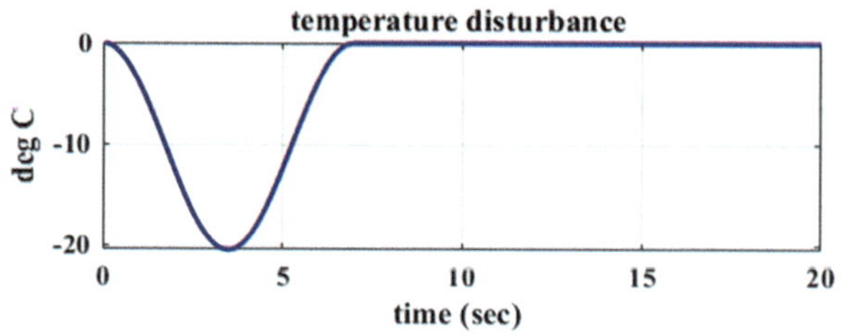

Fig. 5.9 Strip temperature disturbance at mill entry, with mill fully threaded

ultimate increase in the mass flow in the strip between the two adjacent mill stands. This could result in an eventual cobble.

To evaluate the capability of the SDRE-based control, a temperature disturbance at the mill entry was assumed to be tracked such that it caused a phase change between stands 5 and 6, and then between stands 6 and 7, as in Fig. 5.9.[19]

As previously described, the material of the strip was austenite with an equivalent carbon of 0.3% (weight). However, to simulate the change from austenite to ferrite, Shida's relationship[20] was used to simulate a strip with an especially low equivalent carbon content to about 0.027%. Thus when the temperature change happens, the change of phase occurs. The characteristics of the strip then are depicted as in Fig. 5.10.[21]

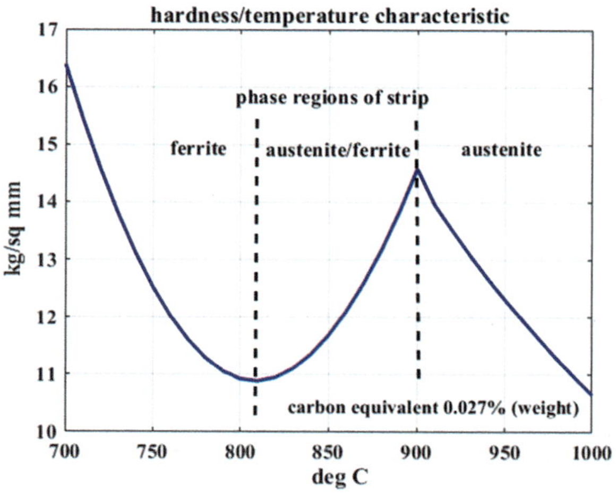

Fig. 5.10 Strip hardness/temperature characteristic at stands 5, 6 and 6, 7

[19] The disturbance is typical. However, due to a lack of available data, adjustments were applied to the disturbance to better simulate conditions that could result in phase changes.

[20] Shida's relationship is described in Chap. 2. The necessary conditions therein are assumed to be implemented to realize the low equivalent carbon content.

[21] The description of the strip conforms to the laboratory measurements of resistance to deformation versus temperature, as described in [14]; a typical example is in Section "Resistance to Deformation Characteristic" of Appendix.

A cobble can result from a difference in the mass flow of the strip between adjacent mill stands. The mass flow is the product of the strip speed and the strip thickness, assuming a nearly constant strip width. Thus a difference in the strip speed at the exit of the upstream stand and the entry speed of the downstream stand can result in a cobble, as the position of the looper depends heavily on the strip speeds of the adjacent stands. This difference can be observed in the position of the looper.

For example, assuming a reasonably tight strip interface between the strip and the looper, an increase in the exit strip speed of the upstream stand, or a decrease in entry strip speed of the downstream stand, will cause an increase in the looper position. This is easily seen in (5.32) wherein the length of the strip $L_{i,i+1}$ depends of the strip speeds at the two stands:

$$L_{i,i+1} = L_{\text{init}} + \int (V_i - V_{i+1})\mathrm{d}t, \tag{5.32}$$

and L_{init}, V_{i+1}, and V_i are the initial length of the strip and the strip speeds of the adjacent stands.

As noted previously, the Q and R matrices (5.21) of the SDRE technique are constant diagonal matrices. In the case of the Q matrix, by setting certain elements of the matrix, a particular variable can be emphasized (i.e., "penalized") during the application. In this instance, the position of the looper is emphasized, with the inherent MIMO control nature of the system suitably controlling other variables to preclude undesirable excursions. This includes the strip speeds at the entry and exit adjacent mill stands, plus others. Some examples are depicted in the following.

Simulations were performed to verify the performance of the control. The operating point of the mill was as depicted in Table 5.1. A disturbance (Fig. 5.9) in the strip temperature at the mill entry was applied, with the mill being fully threaded. The disturbance was modified to include a factor that could result in a phase change as it moved through stand 6, and then adjusted to similarly result in a phase change at stand 7. Figures 5.11 and 5.13 show the changes in hardness as the disturbance in temperature moves through stands 6 and 7. Mass flows at stands 5, 6 and 6, 7 are depicted in Figs. 5.12 and 5.14.

The excursions in the mass flows at stands 5, 6, and 7 were quite small, less than 0.1%. Similarly the deviations in other variables, e.g., looper position, strip thickness, tension, etc., were similarly of the same order of magnitude. However, it was expected that the deviations using different materials might be somewhat greater. Thus simulations were repeated with a lower hardness of about 8 kg/mm^2 (Fig. 5.15). The results were similar (Fig. 5.16), with only slightly larger deviations.

As described previously, the capabilities of the SDRE method provided a useful example in handling a challenging control issue. Additional characteristics and applications are as described further in the following sections.

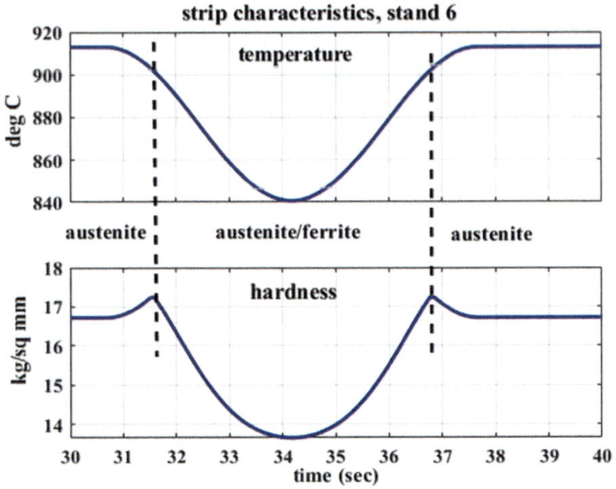

Fig. 5.11 Strip hardness/temperature characteristic at stand 6

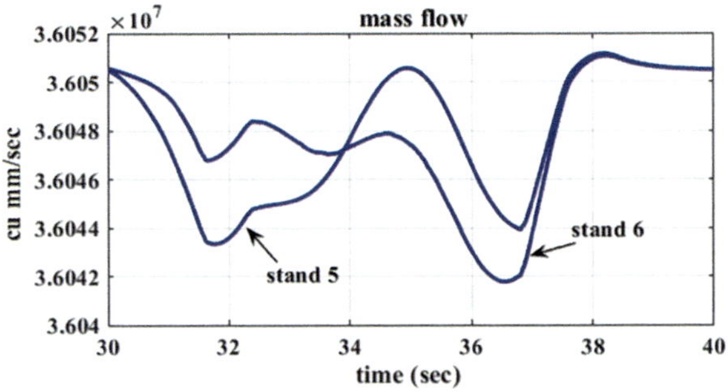

Fig. 5.12 Mass flows at stands 5, 6

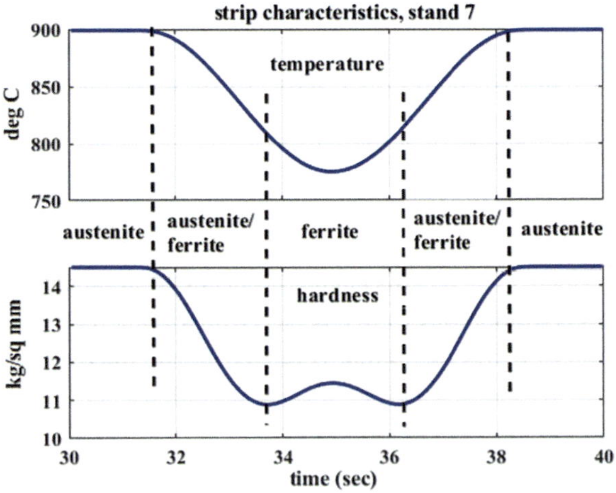

Fig. 5.13 Strip hardness/temperature characteristic at stand 7

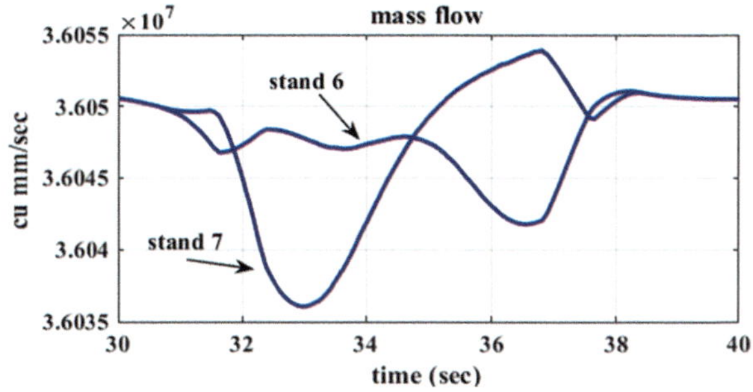

Fig. 5.14 Mass flows at stands 6, 7

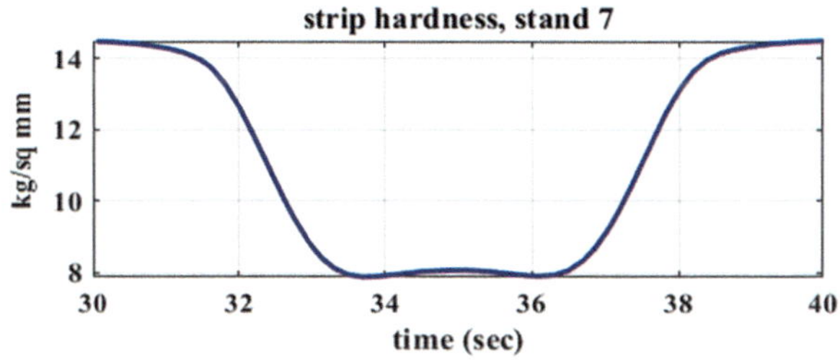

Fig. 5.15 Strip hardness characteristic, with lower hardness, at stand 7

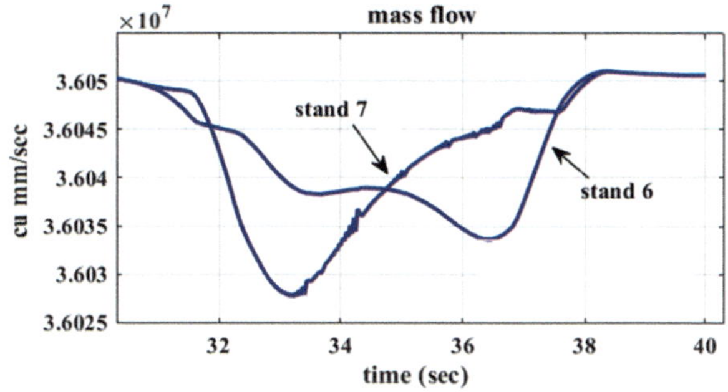

Fig. 5.16 Mass flows at stands 6, 7, with lower hardness at stand 7

5.2.7 *Virtual Rolling for Control of Threading*[22]

Previously, mostly what was described assumed that the mill was fully threaded, with the looper at its operating position. As also noted, the looper plays a significant roll in maintaining the mass flow and the tension of the strip between adjacent mill stands. In the looper, an arm and pivot arrangement provides force to the moving strip. The force is applied by a hydraulic cylinder, or occasionally by a torque motor.

---

[22] Copyright IEEE 2016, figures and tables as shown in this section. Portions of these are reprinted with cosmetic alterations, by permission, from [15]. Unless otherwise noted, data depicted are unchanged from [15].

Fig. 5.17 Threading sequence, stands i, $i + 1$

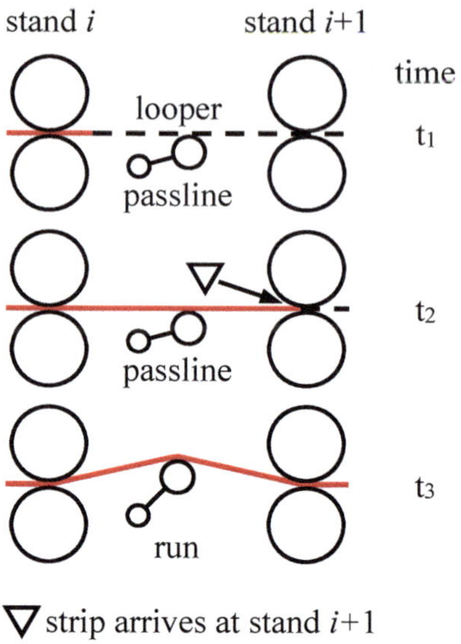

In the threading process, as depicted in Fig. 5.17, the head of the strip is moved from stand to stand until the entire mill is threaded. At time t_1 the head of the strip has been threaded into stand i and is moving toward stand $i + 1$. The looper is resting at the passline position. At time t_2 the head of the strip has just arrived at (and is in) stand $i + 1$, with the looper just touching the strip. At time t_3 tension in the strip is applied, and the looper is raised to its run position for the threading of the remainder of the mill.

As might be expected, as the head of the strip moves through the mill the mathematical model of the process changes considerably. This presents a significant challenge to the control, wherein the action of the looper plays a major part. To enhance the control performance a virtual technique was developed and applied to the SDRE to simulate the rolling function of the unthreaded portion of the mill.

During threading, in addition to handling the changes in the process model and its dynamics, the controller should preclude excessive excursions in strip tension and looper position when the looper is raised from passline to run position; and likewise during the fast drop in the work roll speed when the head end of the strip enters a mill stand.[23] These undesirable excursions have been significantly reduced as will be seen in what follows.

The previously described SDRE controller has been modified to improve the handling of the threading phase. The updated controller is depicted in Fig. 5.18.

In this upgraded control the unthreaded portions of the mill are simulated by a model in the controller to produce a virtual rolling condition. Further, logic is added so that by combining the signals of the virtual rolling and the actual rolling, the

[23] There has been a need for improvement in these areas, as noted in [16] and in the related references cited therein.

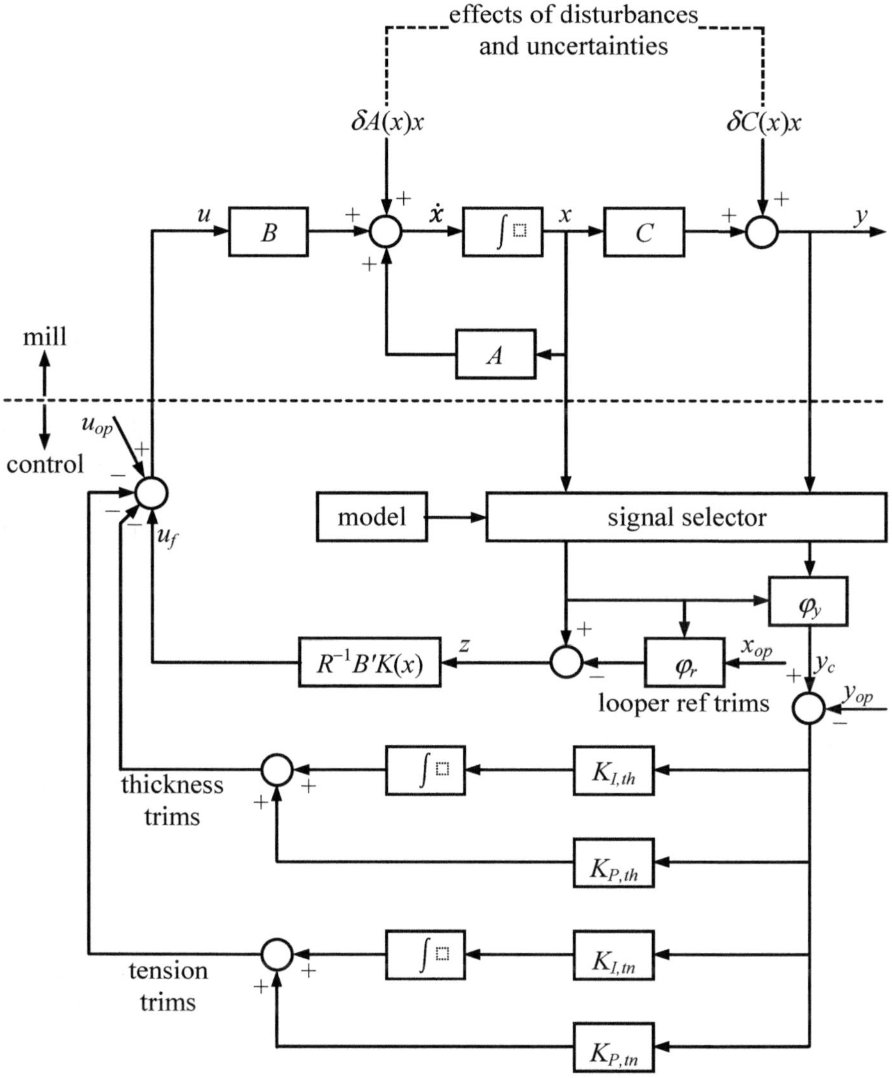

Fig. 5.18 SDRE hot mill control with upgrades for virtual rolling

controller sees during threading, the process as a fully threaded mill. Thus, in this method, the control loops are closed and fully operational during threading.

In this method, actual measurements are used for work roll speeds, looper positions and torques, and positions of the hydraulic cylinders for the roll bites. Simulated measurements are used for strip tensions, roll forces, and strip thicknesses. Strip temperatures and equivalent carbon contents are also simulated, although these generally are not measured during the actual threaded rolling.

In the movement of the strip through the mill during threading, the values of the simulated variables are automatically switched to the values of the measured variables. Thus the interactions between the controller and the virtual portions are switched to interactions between the controller and the actual portions as the strip continues in the threading function. The differences between the virtual and the actual variables are less significant since the model used to determine the virtual variables is based on extensive collections of data from previous applications, considerable experience, plus methods applied, mostly in level 2, to refine measurements and models to more closely represent present actual values. Thus a simplification in control is realized over previous methods that must consider significant changes in the models during the threading sequence.

The status of certain functions before an incoming bar enters the mill, which is in virtual mode, are as follows. Generally, measurements or estimates requiring the actual strip in the mill are in the virtual mode. Certain others can be in the run mode, with references based on the estimated characteristics of the upcoming strip.

Some examples of being in the run mode are

- looper positions, with references that are set to control the loopers to be at passline positions;
- work roll speeds;
- hydraulic cylinder positions;
- others as needed.

Some examples of being in the virtual mode, which then are switched to being in the run mode when the strip is detected as being in the first or second mill stands are

- the measured roll force of stand 1;
- the estimated strip thickness at the exit of stand 1;
- the measured roll force of stand 2;
- the estimated strip thickness at the exit of stand 2;
- the measured strip tension as the strip enters (and is in) stand 2;
- others as needed.

When the head of the strip is in stand 2, the tension reference is switched to run tension, which establishes run tension in the strip. The reference for looper position is already in run mode, and with tension established, the looper then is raised to run position. Functions between stands 1 and 2 are then in run mode, and the head of the strip moves to stand 3 to repeat the sequence.

Some of the advantages of what was previously described include the improvement in the reduction of undesirable deviations in the interactions of the various variables throughout the mill. This is largely due to the nonlinear nature of the MIMO controller, the model, and the control during both threading and running, as these better represent the actual process.

As an example, retaining closed control loops during the virtual and run modes significantly reduces the peak of the looper position during the raising of the looper from passline to run position, which improves the interactions between the looper and strip. Additionally, the disturbance due to the impact drop in speed when the

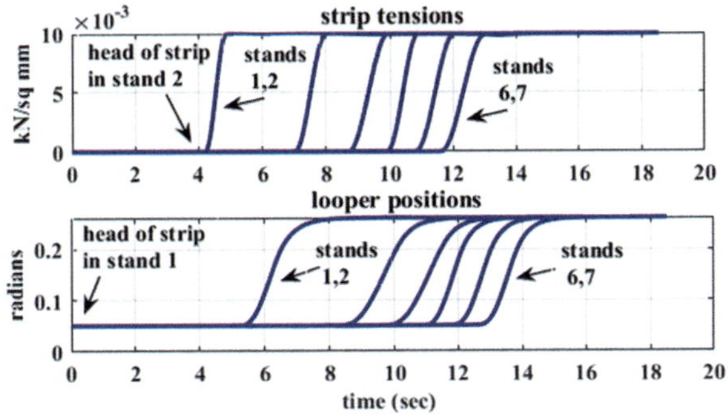

Fig. 5.19 Strip tensions and looper positions during threading, without disturbances or uncertainties

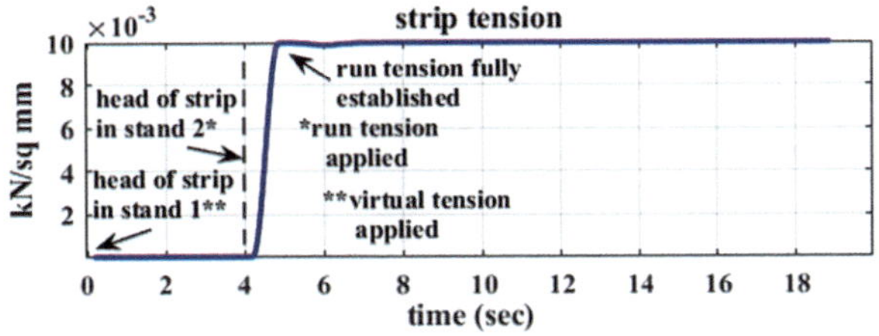

Fig. 5.20 Strip tension (stands 1, 2) during threading showing response of tension during switch from virtual to run

head of the strip enters the roll bite of the downstream stand is considered. More specific detail of these is provided in what follows.

To gain some better insight into the basic controller design, simulations were applied initially without uncertainties or disturbances. Then uncertainties were applied to simulate mismatches likely to occur during switching from virtual to run modes, with comparison to other methods. Initially the looper between stands 1 and 2 was considered when raising from passline to run positions. For comparison, the looper position and strip tension are depicted initially without uncertainties or disturbances, such as the impact drop in speed at the downstream stand. The results for the looper between stands 1, 2 are shown in Figs. 5.19, 5.20 and 5.21. The deviations in tensions between other stands are between – 0.5% (stands 3, 4) and – 1.6% (stands 6, 7). A general comparison with other methods is made in Table 5.4.

Additional work was done to evaluate performance with uncertainties in the measurements of tension and roll force, the estimate of strip thickness at a stand exit, the estimates of temperature and equivalent carbon content of strip at the mill entry, and with the disturbance of the speed impact drop as the strip enters a mill stand. The mill area between stands 1 and 2 was considered, which was taken as reasonably typical for most areas in the remainder of the mill. The characteristics of the uncertainties are as follows:

- in the case of the tension, the uncertainty was taken to be ± 20% of measured tension with the looper at passline, and ± 5% with the looper at run position;
- in the case of the roll force measurement, an uncertainty of ± 10% was taken, which included the uncertainty in the force measurement and the uncertainty in

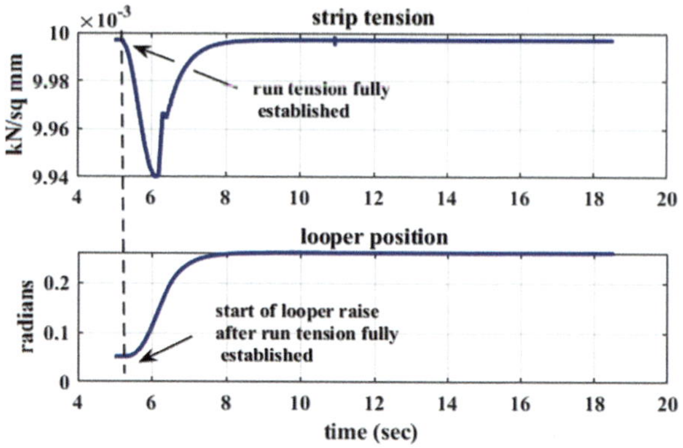

Fig. 5.21 Strip tension and looper position (stands 1, 2) during threading showing responses in tension and looper during raising of looper

Table 5.4 Excursions in tension during raising of looper comparison with other methods

Method	Percent deviation (max) (%)
SDRE herein	0.6
Method 1 [16]	5.0
Method 2 [17]	12.0
Method 3 [18]	30.0
Older methods	50–200

the estimation of the mill modulus; these both affect the estimate of the strip thickness;
- in the cases of the incoming strip temperature and equivalent carbon, the uncertainties were taken as ± 20 °C and ± 20% of the values taken for the virtual conditions.

The uncertainties were divided into the following four groups:

- Group 1: tension + 20%, + 5%; roll force + 10%;
- Group 2: tension + 20%, + 5%; roll force − 10%;
- Group 3: tension − 20%, − 5%; roll force + 10%;
- Group 4: tension − 20%, − 5%; roll force − 10%.

In each group the speed impact drop was included, as were the uncertainties in the temperature and the equivalent carbon content. The uncertainties were applied as step functions, slightly softened, as the head of the strip entered the roll bite. The simulated impact drop in speed (Fig. 5.22) was applied concurrently with the uncertainties. However, deviations due to the uncertainties in temperature and equivalent carbon content initially had lesser effects, and therefore were not separately evaluated in this effort. The results of the simulations are depicted in Figs. 5.23, 5.24, 5.25 and 5.26.

What was presented previously describes the potential for some reasonable improvements in the control of threading. Some of these include the control of the looper position and strip tension during the raising of the looper from passline to run

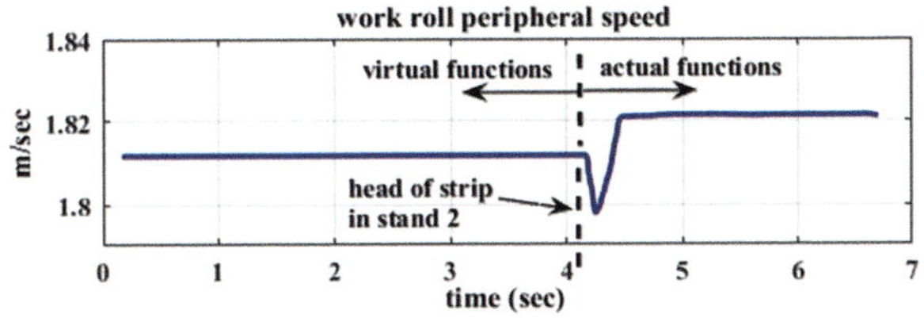

Fig. 5.22 Work roll peripheral speed (stand 2) during threading showing speed impact as strip head enters the stand gap, with disturbance, without uncertainties

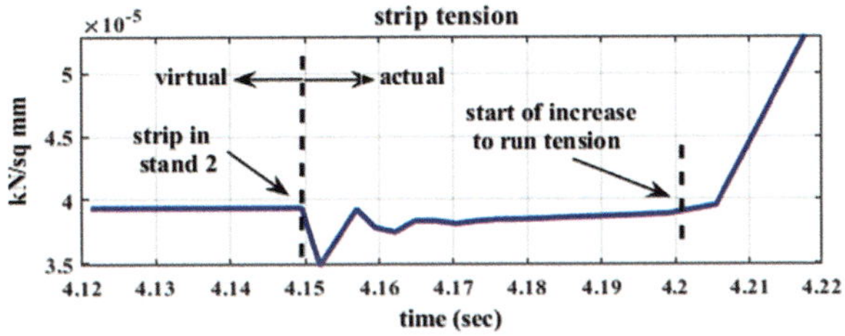

Fig. 5.23 Strip tension (stands 1, 2) during threading showing start of increase in tension during raising of looper, with disturbance, without uncertainties

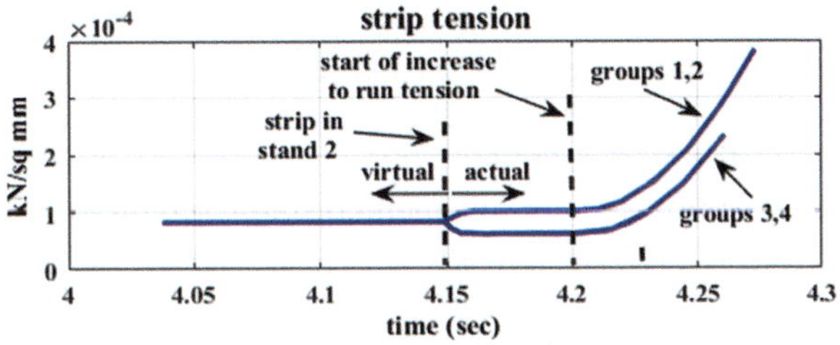

Fig. 5.24 Strip tension (stands 1, 2) during threading showing start of increase in tension during raising of looper, with uncertainties and disturbance (negligible)

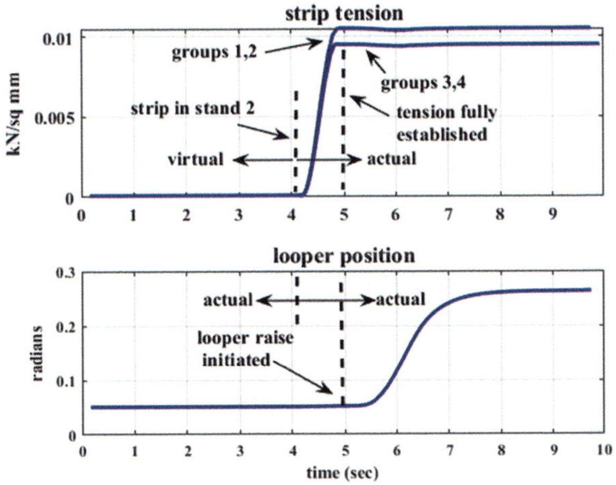

Fig. 5.25 Strip tension and looper position (stands 1, 2) during threading showing responses in tension and looper during raising of looper, with uncertainties and disturbance (negligible)

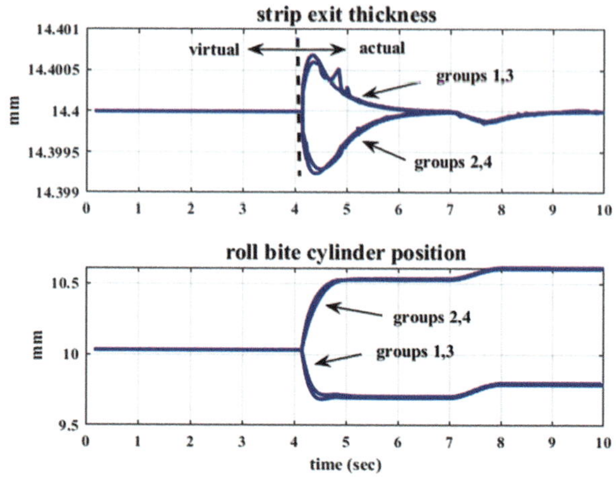

Fig. 5.26 Stand 2 strip exit thickness (estimated) and cylinder position during threading showing responses as strip head exits stand 2, with uncertainties and disturbance (negligible)

position. Previously, the application of other methods resulted in excessive excursions in the looper position and strip tension, which caused undesirable effects in the strip. As noted in the previous figures, these excursions were significantly reduced using the new method.

Additionally, the new method simplified the control as the on-line solution of the algebraic Riccati equation was usually unnecessary, which considerably reduced the complexity of the design process. Also, the inherent matching of the ordinary differential equations between subsequent controller scans during the switch of the control mode from virtual to actual results in improved performance, plus simplification during the threading process. It is considered that further investigation, actual application, and improvement in the uncertainty in models and measurements (mostly in level 2) will result in additional advantages, especially in support of start-up wherein much of the work of matching the controller to the actual process would be even further reduced.

5.2.8 Virtual Rolling for Robustness to Faults[24]

The handling of faults in certain measurements can help improve the availability of the process. Two of the faults that are examined as examples are the measurements of strip tension stress and roll force. Certain faults in either of these could degrade the quality of the mill output and result in a loss of production, plus possible damage to equipment. In what follows shows how the SDRE control coupled with virtual

[24] Copyright IEEE 2017, figures and tables as shown in this section. Portions of these are reprinted with cosmetic alterations, by permission, from [19]. Unless otherwise noted, data depicted are unchanged from [19].

rolling can respond to these faults such that rolling can continue without significant disruption of the process or other undesirable situations.

At each mill stand there are measurements of tension stress and roll force. These measurements are determined by sensors which feed signals to the controller via instrument channels. A fault in the sensor or channel, or other abnormality, can cause a deviation from the desired value by more than a predetermined amount. In either case the controller should be able to handle abnormalities that are fast acting or develop slowly, so that a controller must be robust to suitably handle abnormal conditions. What follows are some basic examples of how the SDRE-based controller with virtual rolling reasonably responds to these two faults.

In the actual controller the following are typical for a fault in the measurement of the tension between mill stands 1 and 2, and in the measurement of the roll force at stand 1. It is assumed that these faults could occur separately or together, with the remainder of the process being without faults. It was considered that the evaluation of these particular faults would provide a reasonable basis for dealing with other faults. This is because the method for dealing with them would be straightforward and similar to other faulted conditions, and thus it would be easier to determine how the mill could be kept running during these conditions. Figure 5.27 presents a functional schematic of the overall operation in the presence of these two faults.

In Fig. 5.27 the variables virtual tension and virtual roll force are generated in the model which is described in Chap. 2. The model is expected to describe the actual process reasonably well as it is based on a wide range of data, experience, and is updated continuously during actual operations for a wide variety of material being processed. Thus the differences between the virtual estimates, as generated in the model, and the actual process as determined by measurements, are minor under normal operating conditions. The interfacing signals are from devices that have a

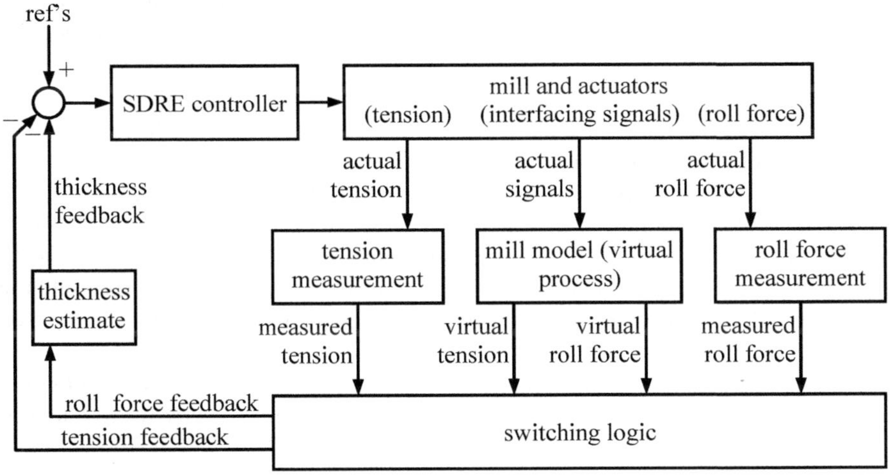

Fig. 5.27 Functional schematic for fault estimation

Table 5.5 Uncertainties

Function	Percent of actual value (%)
Tension, actual	± 5
Tension, virtual	± 10
Roll force, actual	± 5
Roll force, virtual	± 10

Table 5.6 Operating ranges[25]

Function	Reference value	Operating range
Tension, kN/mm^2	0.01	0.0081–0.0121
Roll force ref gain	1.00	0.06–1.35

close interaction with the strip during actual operation. For example, some of these are local controllers, actuators, and sensors, such as for looper torques and positions, work roll speeds, and roll bite cylinder positions.

In the case of the tension, when the measurement is outside an acceptable operating range, a switch from measured to virtual is done. The switching logic switches the feedback from measured to virtual. However, the virtual tension can have a slightly greater uncertainty than an unfaulty actual tension. This could allow the continued processing of the strip, but with possibly some degradation in final quality, but still be useful in a limited sense, and avoid a possible cobble with its undesirable effects.

In the case of the roll force, the switching of the roll force from measured to virtual is determined by the characteristics of the product being processed, especially its resistance to deformation (i.e., hardness). This occurs as the desired roll force is determined by various operational situations, such as changes in the resistance to deformation in the strip being processed. To account for this, if a gain, as determined by the ratio of measured roll force to virtual roll force, is beyond a specified range, a switch to virtual roll force is made. More detail is provided in the results of simulations which follow.

Uncertainties were considered to examine control performance during various situations in operation. The uncertainties are as in Table 5.5. Disturbances were handled well by the controller, as shown in the simulations.

The acceptable operating ranges of tension and roll force are as in Table 5.6.

The controller structure is similar to that of Sect. 5.2.7, except that a function is added as depicted in Fig. 5.28 to include the features of Fig. 5.27.

To help get some initial insight to fault tolerance, simulations were done without disturbances or uncertainties to confirm the basic concepts. The results are depicted in Figs. 5.29, 5.30, 5.31 and 5.32.

The previous evaluation was then expanded to include uncertainties. In this simulation a measurement uncertainty occurred at 5 s, with a measurement fault at 15 s,

[25] Operating ranges include uncertainties.

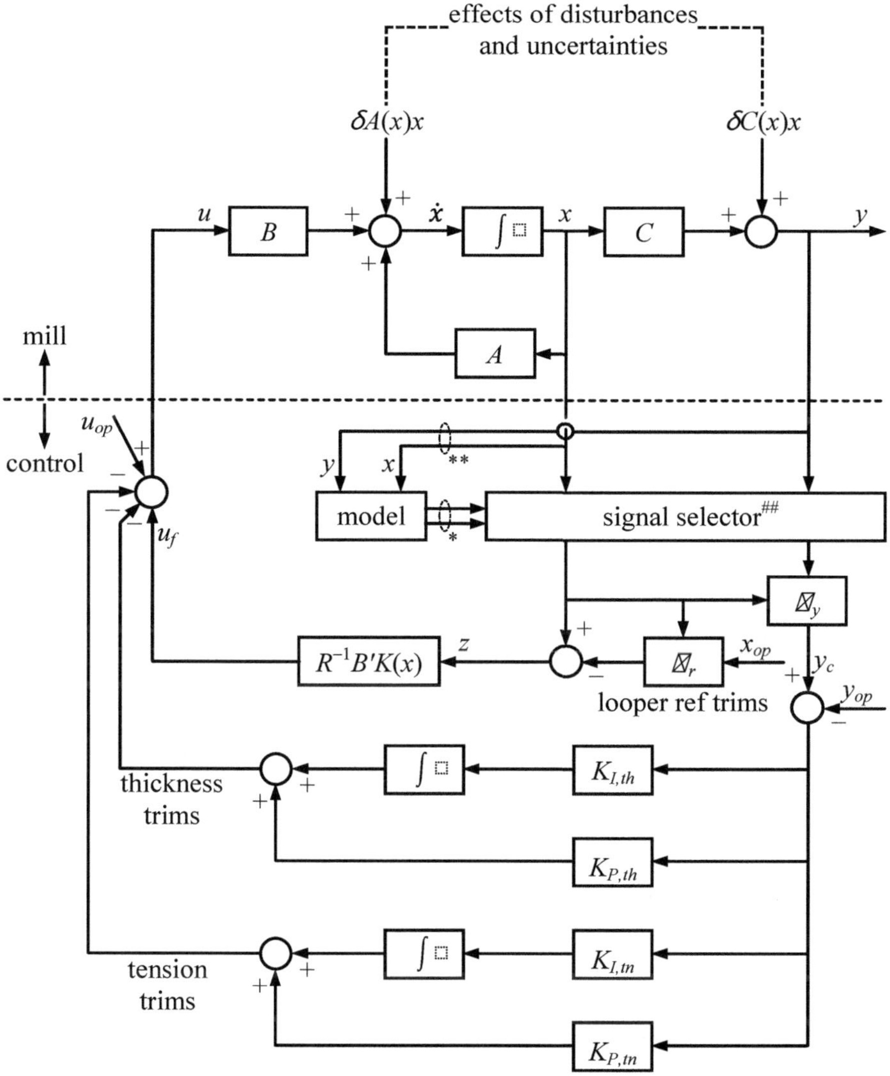

Fig. 5.28 SDRE hot mill control with upgrades for virtual rolling and improved fault tolerant techniques. Improvements for fault tolerance: ## signal selector includes switching logic (Fig. 5.27). * Model provides virtual signals for tension and roll force. ** Inputs to model for interfacing signals (Fig. 5.27)

and with a subsequent switch from the measured to the virtual. The results are as depicted in Figs. 5.33, 5.34 and 5.35.

Figure 5.36 depicts the responses in the roll force resulting from disturbances in the entry thickness at stand 1. As noted, the deviations are not excessive in either the actual or virtual measurement of the roll force. In Figs. 5.37 and 5.38 a fault in the

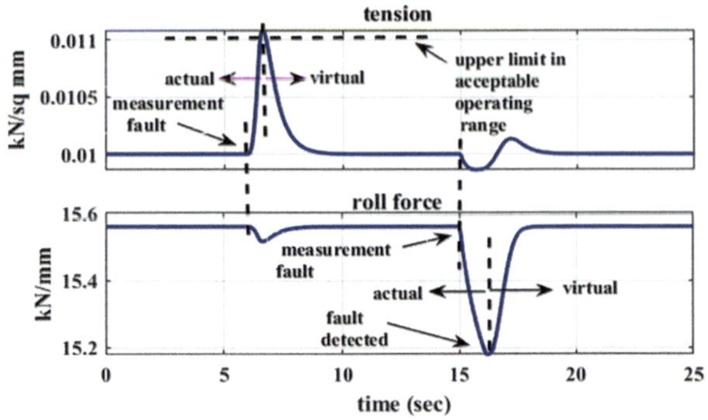

Fig. 5.29 Responses to faults in tension stress (stands 1, 2) and roll force (stand 1), without uncertainties or disturbances

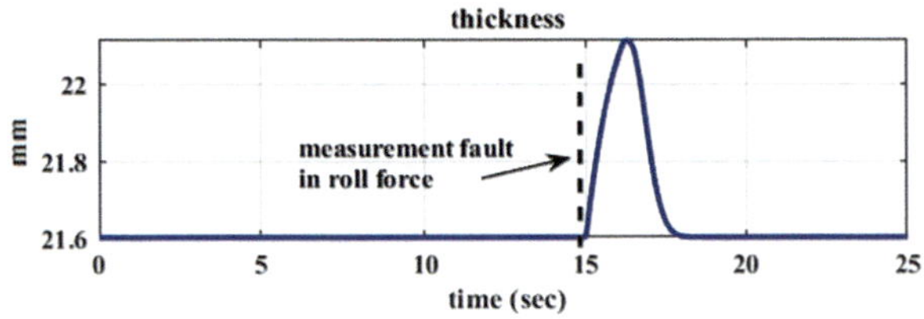

Fig. 5.30 Effect on strip exit thickness (stand 1) by fault in the roll force as in Fig. 5.29

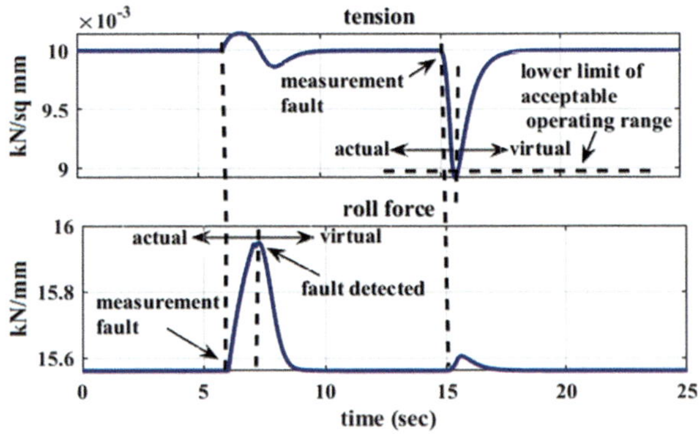

Fig. 5.31 Responses to faults in tension stress (stands 1, 2) and roll force (stand 1), without uncertainties or disturbances, with fault as in Fig. 5.29

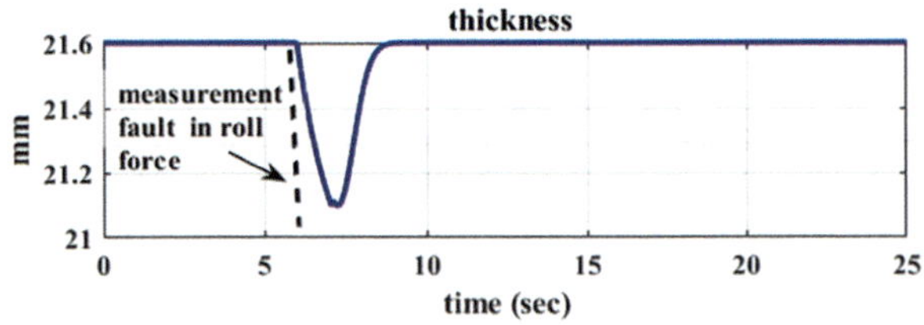

Fig. 5.32 Effect on strip exit thickness (stand 1) by fault in the roll force as in Fig. 5.31

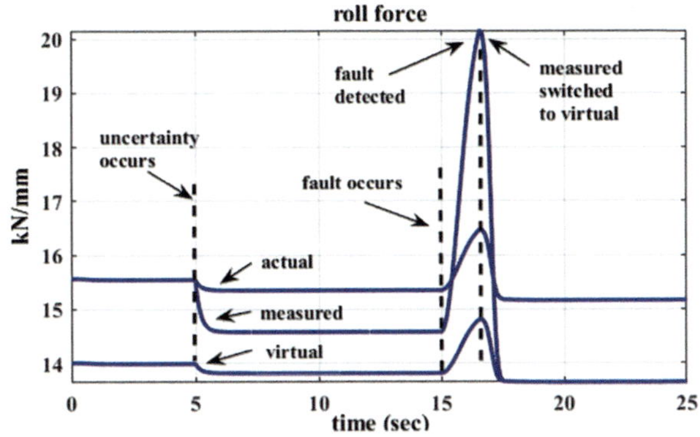

Fig. 5.33 Effects of uncertainty and fault in the roll force measurement (stand 1), with a switch from measured to virtual

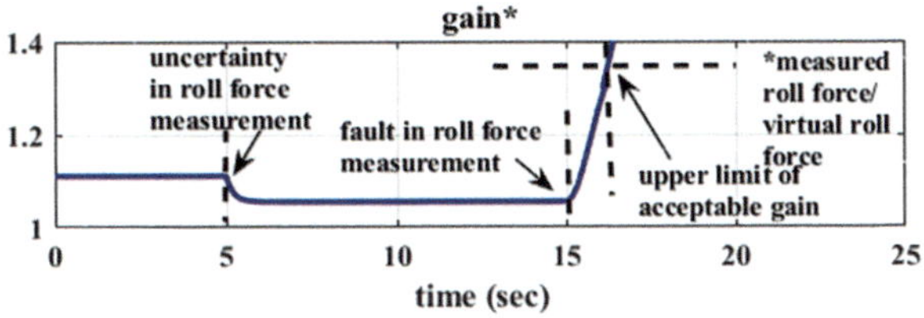

Fig. 5.34 Partial detail of response of roll force in Fig. 5.33

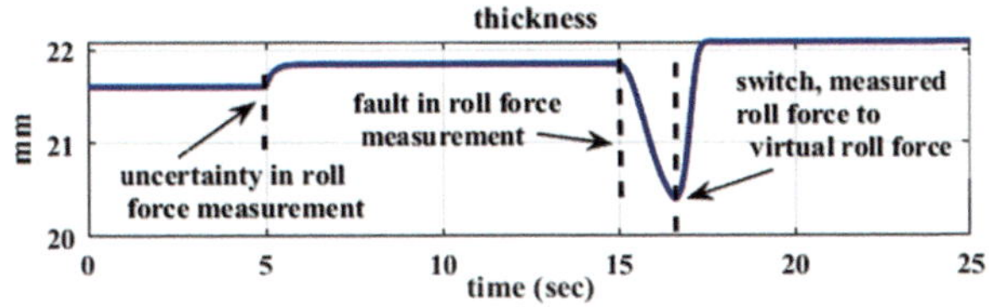

Fig. 5.35 Response of the strip thickness at the exit of stand 1 due to uncertainty and fault in roll force measurement as in Fig. 5.33

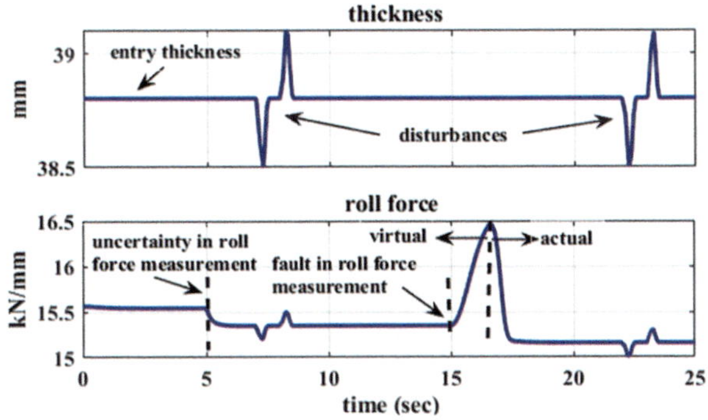

Fig. 5.36 Response of the roll force at stand 1 due to uncertainty and fault in the roll force measurement, with disturbances in the entry thickness

tension measurement occurs with the switch to virtual tension as the fault is detected. In Fig. 5.39 it is shown that disturbances in looper torque have minor effect on the looper tension.

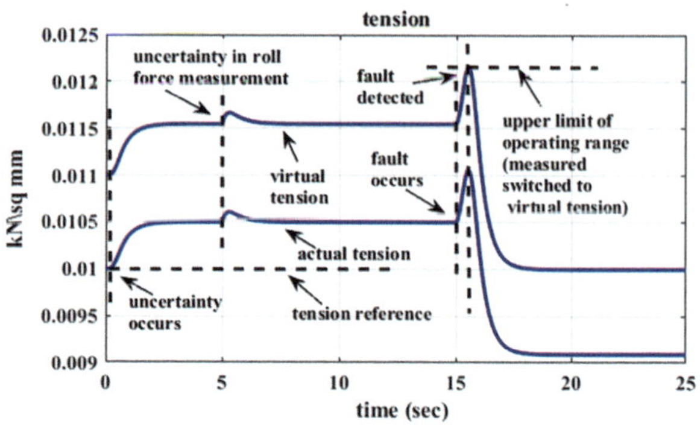

Fig. 5.37 Response of the tension stress (stands 1, 2) due to uncertainty and fault in the tension measurement, and with an uncertainty in the roll force measurement

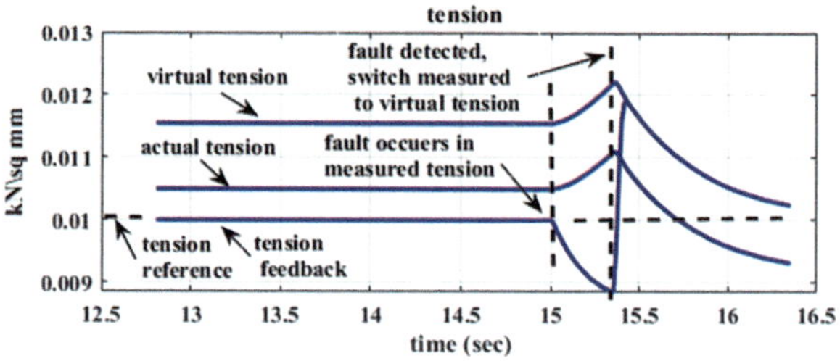

Fig. 5.38 Detail of transfer of tension feedback (stands 1, 2) as depicted in Fig. 5.37 in response to a fault in tension measurement

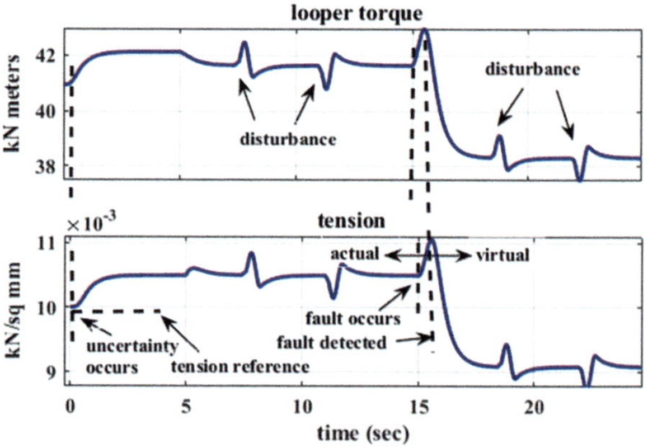

Fig. 5.39 Effect of disturbances in looper torque on strip tension

The previous plots have shown that the system is reasonably effective in handling certain disturbances and uncertainties. Of course, not every combination of these can be considered. However, what has been depicted gives some basic insight into the capability of maintaining the operation of the process to provide a product that can retain some useful characteristics and reduce the likelihood of major problems such as cobbles.

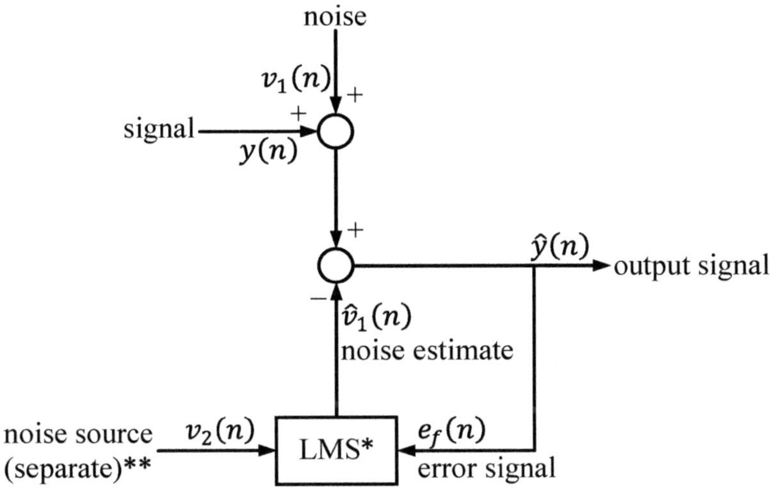

Fig. 5.40 Adaptive noise cancelation using a least mean square filter. * Least mean square adaptive filter. ** Correlated with noise

5.2.9 Eccentricity Compensation

A method of compensation for the eccentricity of mill rolls that fits in well with SDRE control is described in what follows.[26] In this technique the relationship (3.2) for the strip output thickness at a mill stand is modified slightly by adding a term for the eccentricity as:

$$h_{\text{out}} = S + S_0 + \frac{F}{M} + e, \tag{5.33}$$

where e is the eccentricity. The method of compensation depends on the characteristic of the backup roll eccentricity being periodic and with a basic frequency proportional to the angular velocity of the roll, i.e., there is a correlation between the eccentricity and the roll's angular velocity.

The basis for this technique is noise cancelation, which depends on the correlation of a noisy signal and some measured noise generated by a separate source. This concept is well-known and is presented in several texts, e.g., [21, 22]. The basic function is depicted in Fig. 5.40.

In Fig. 5.40, the discrete time step is represented by n, $v_1(n)$ and $v_2(n)$ are correlated noise sources, and signal $y(n)$ is uncorrelated with signals $v_1(n)$, $v_2(n)$, and $\hat{v}_1(n)$,

[26] The method is based on the technique described in [8] and its references. It can be noted that many other recent methods have been proposed and are described in various papers, e.g., [20], etc. However, in almost every instance the eccentricity can be reasonably approximated as a series of sinusoids. Thus, it is felt that what is presented herein is relatively straightforward and can help provide some basic understanding into eccentricity and its control.

where $\hat{v}_1(n)$ is the output of the LMS (least mean square) adaptive filter. Additionally it is assumed that $v_2(n)$ contains no components of $y(n)$.

Using Fig. 5.40, the following is applicable:

$$\hat{y}(n) = y(n) + (v_1(n) - \hat{v}_1(n)), \tag{5.34}$$

and

$$e_f(n) = \hat{y}(n), \tag{5.35}$$

where the error signal applied to the filter is $e_f(n)$.

Now squaring and taking expected values:

$$E\left[e_f^2(n)\right] = E\left[\hat{y}^2(n)\right], \tag{5.36}$$

where

$$\hat{y}^2(n) = y^2(n) + 2y(n)(v_1(n) - \hat{v}_1(n)) + (v_1(n) - \hat{v}_1(n))^2. \tag{5.37}$$

Since $y(n)$ is not correlated with $v_1(n)$ or $\hat{v}_1(n)$,

$$2E[y(n)(v_1(n) - \hat{v}_1(n))] = 0. \tag{5.38}$$

The LMS filter thus minimizes $E\left[e_f^2(n)\right]$ and thus minimizing $E\left[(v_1(n) - \hat{v}_1(n))^2\right]$ but not affecting $y(n)$, so that the noise in the output signal $\hat{y}(n)$ is reduced. An example of this technique as applied to control of eccentricity in a tandem rolling mill is depicted functionally in Fig. 5.41.

In Fig. 5.41, r represents the cylinder position reference, y is the position feedback, w is a measurement whose frequency is proportional to the angular velocity of the backup roll, d represents the disturbance that describes the backup roll eccentricity, Y_{LMS} and U_{LMS} are the inputs and outputs of the LMS filter, C_u is a transfer function describing an input–output linearization of the nonlinear controller; K_L, R_L, C_d, and Q_d are transfer functions, where Q_d is determined such that $|Q_d(j\omega)\, C_u(j\omega) - 1|$ is nearly zero in the range of frequencies characteristic of the eccentricity. The overall main idea is that the disturbance $C_d\, d$ is cancelled by $C_u\, Q_d\, U_{LMS}$, which results in an output y to be essentially eccentricity free, with significantly reduced effects of eccentricity in the strip.

Additional insight into the functioning of this method is realized by comparison of the signals in Fig. 5.40. In Fig. 5.41 the signals Y_{LMS}, $C_u\, Q_d\, U_{LMS}$, and $C_d\, d$ correspond to $e_f(n)$, $\hat{v}_1(n)$, and $v_1(n)$, respectively, in Fig. 5.40. Also, in Fig. 5.40 $y(n)$ is taken as zero, and a sinusoid of frequency ω is generated to correspond to $v_2(n)$. Thus the estimated eccentricity signal $C_u\, Q_d\, U_{LMS}$, corresponding to the estimated noise signal $\hat{v}_1(n)$, is subtracted from the eccentricity signal $C_d\, d$, corresponding to the noise signal $v_1(n)$, and thus reduces the eccentricity component in the output y.

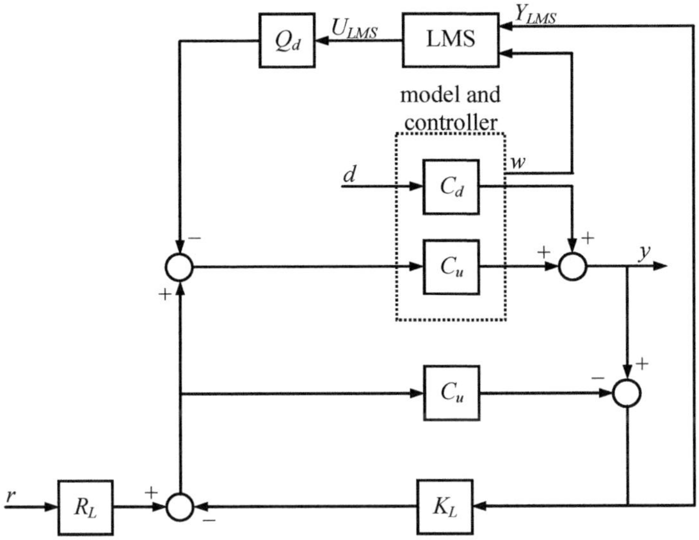

Fig. 5.41 Cylinder position controller with eccentricity compensation

Further, it is assumed that any eccentricity noise in the signals representing the cylinder position or the roll force is negligible by being suitably filtered. The effects on deviations in the roundness of the backup roll, the effects of the work roll, the other backup roll, plus others that are recognized as harmonics in the eccentricity signal, are sinusoidal and therefore can be handled appropriately. Examples are simulated as described in Section "Examples of Eccentricity Compensation" of Appendix. Thus as shown in this brief evaluation, eccentricity compensation is certainly reasonable for use with the SDRE controller.

5.2.10 *Cost Reduction and Energy Saving*[27]

In general, the yield of the tandem hot rolling process has been high. However, there still are losses resulting due to product that is out of tolerance, in returns from ultimate users, and others. The use of the SDRE technique offers a strong potential for further improvement in the quality and yield which can reduce costs and save energy, with a subsequent realization of an increase in profits. Additionally, the reduction of scrap produced also reduces the energy consumed by reprocessing via other areas such as electric arc furnaces and subsequent functions. Some examples are as follows.

[27] Copyright IEEE 2013, figures and tables shown in this section. Portions of these are reprinted with cosmetic alterations, by permission, from [23]. Unless otherwise noted, data depicted are unchanged from [23].

At present approximately 90 million tons of steel per year is produced domestically, about half of which is hot rolled coils. With a current price of about 800 USD/ton, and an estimated yield loss of roughly 1.5%, the total yield loss is about one half billion USD/year.[28] Thus any improvement in control, even if appeared to be minor, could reduce yield loss and reduce costs significantly.

Additionally, the product loss in this hot rolling process is about 675,000 tons per year, as mostly scrap. The reprocessing of the scrap in electric arc furnaces, continuous casters, and hot rolling mills would require about 285 kWh of energy. Thus almost any reasonable improvement in control would help in the saving of energy.

Simulations were performed earlier herein to estimate controller capabilities under certain normal and abnormal conditions. Some of these include disturbances and uncertainties that might occur during normal operating situations. Certain others, for example, are those described in Sect. 5.2.6 wherein abnormal strip temperature excursions caused undesirable characteristics in the strip material that have a good chance of resulting in a cobble. In many instances the performance of the SDRE-based controller under these various conditions significantly exceeded that of most conventional methods.

In the following simulations the uncertainties considered were large variations in unmeasured variables applied concurrently with disturbances in the incoming strip.[29] Additionally a typical zoom in the mill speed was also applied concurrently. The unmeasured variables were strip equivalent carbon content, temperature, thickness, and mill modulus. The unmeasured disturbances were strip temperature and thickness. The applications were applied in groups as described in Table 5.7 and Fig. 5.42.

Table 5.7 Uncertainties and disturbances

Group	Eq. carb. (%)		Temp. (°C)		Thick (%)		Mill mod. (%)	
	Uncert.	Disturb.	Uncert.	Disturb.	Uncert.	Disturb.	Uncert.	Disturb.
1	– 70	0	– 30	– 25	20	5	20	0
2	– 70	0	– 30	– 25	20	5	– 20	0
3	– 70	0	– 30	– 25	– 20	– 5	– 20	0
4	– 70	0	– 30	– 25	– 20	– 5	20	0
5	70	0	30	25	20	5	20	0
6	70	0	30	25	20	5	– 20	0
7	70	0	30	25	– 20	– 5	– 20	0
8	70	0	30	25	– 20	– 5	20	0

[28] The data for these estimates are derived from [24, 25].

[29] Uncertainties and disturbances in measured variables were less severe; their effects were addressed in previous sections. An exception is the entries for $h_{out,7}$; these are based on a measurement at the mill exit.

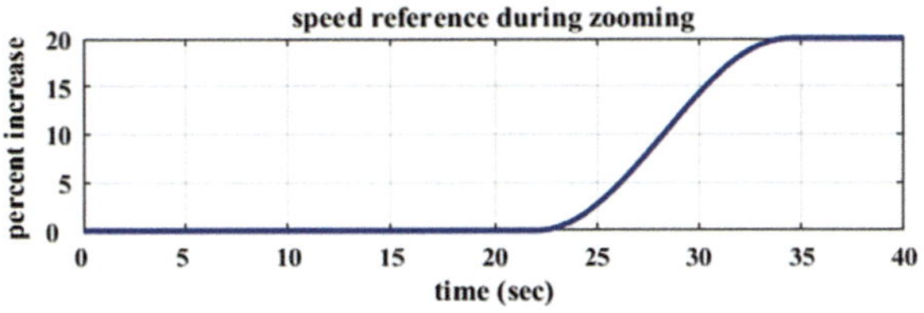

Fig. 5.42 Increase in mill speed reference during zooming

Fig. 5.43 Group 1, strip temperature and thickness at mill entry, with uncertainties and disturbances

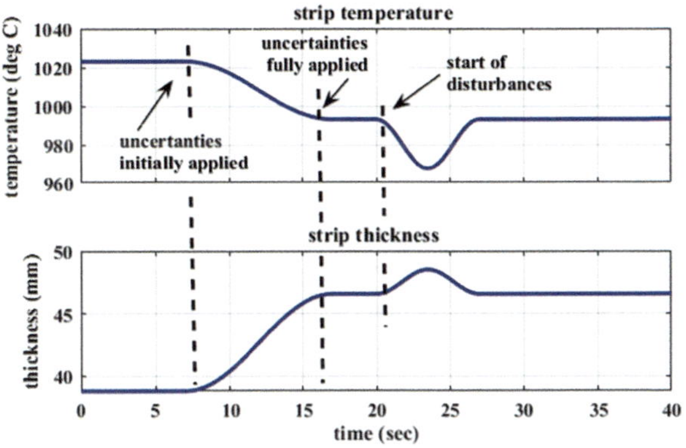

Fig. 5.44 Group 1, stand 6, 7 tension stress and looper position, and stand 7 output thickness, in response to perturbations depicted in Fig. 5.43

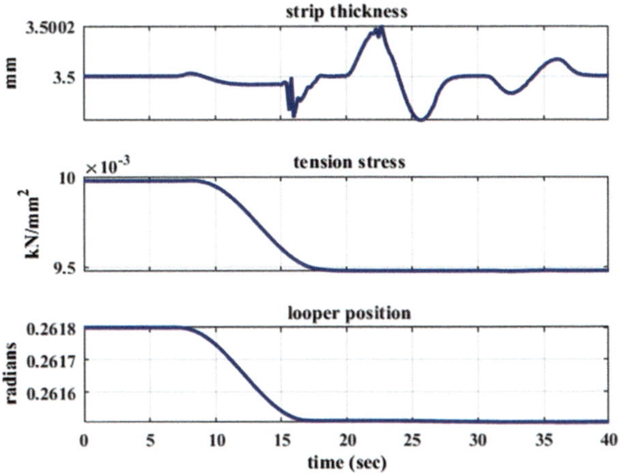

The deviations in operating point values resulting from these uncertainties and disturbances are depicted in Tables 5.8, 5.9, 5.10, 5.11, 5.12, 5.13, 5.14 and 5.15,[30] and as an example for Group 1 in Figs. 5.43 and 5.44.

[30] The notation for variables is as follows: strip thickness (h) and tension stress (σ), looper position (θ), and strip temperature (T).

Table 5.8 Group 1 responses

Variables	Mean	Maximum deviation	Standard deviation
$h_{out,1}$–$h_{out,6}$	3.65%	− 5.78	1.11
$h_{out,7}$	0	0.004	0.002
σ_1–σ_6	− 5.07	− 5.26	0.009
θ_1–θ_6	− 0.01	− 0.11	0.004
T_1–T_6	− 3.53	− 5.75	0.77
T_7	− 3.78	− 5.83	0.80

Table 5.9 Group 2 responses

Variables	Mean	Maximum deviation	Standard deviation
$h_{out,1}$–$h_{out,6}$	5.78%	8.43	1.37
$h_{out,7}$	0	0.006	0.003
σ_1–σ_6	− 5.07	− 5.29	0.010
θ_1–θ_6	− 0.093	− 0.10	0.009
T_1–T_6	− 3.53	− 5.75	0.77
T_7	− 3.58	− 5.83	0.80

Table 5.10 Group 3 responses

Variables	Mean	Maximum deviation	Standard deviation
$h_{out,1}$–$h_{out,6}$	5.14%	8.33	1.72
$h_{out,7}$	0	− 0.006	0.003
σ_1–σ_6	− 5.07	− 5.30	0.09
θ_1–θ_6	− 0.094	− 0.10	0.009
T_1–T_6	− 3.53	− 5.75	0.77
T_7	− 3.78	− 5.83	0.80

Table 5.11 Group 4 responses

Variables	Mean	Maximum deviation	Standard deviation
$h_{out,1}$–$h_{out,6}$	− 3.29%	− 5.74	1.15
$h_{out,7}$	0	0.002	0.001
σ_1–σ_6	− 5.07	− 5.26	0.09
θ_1–θ_6	− 0.10	− 0.11	0.004
T_1–T_6	− 3.53	− 5.75	0.77
T_7	− 3.78	− 5.83	0.80

Table 5.12 Group 5 responses

Variables	Mean	Maximum deviation	Standard deviation
$h_{out,1}$–$h_{out,6}$	– 3.40%	– 5.26	0.96
$h_{out,7}$	0	0.002	0.001
σ_1–σ_6	– 5.07	– 5.26	0.023
θ_1–θ_6	– 0.010	– 0.11	0.004
T_1–T_6	3.32	5.43	0.74
T_7	3.68	5.83	0.73

Table 5.13 Group 6 responses

Variables	Mean	Maximum deviation	Standard deviation
$h_{out,1}$–$h_{out,6}$	5.41%	7.67	1.26
$h_{out,7}$	0	– 0.005	0.002
σ_1–σ_6	– 5.08	– 5.28	0.096
θ_1–θ_6	– 0.093	– 0.10	0.009
T_1–T_6	3.32	5.43	0.74
T_7	3.68	5.83	0.73

Table 5.14 Group 7 responses

Variables	Mean	Maximum deviation	Standard deviation
$h_{out,1}$–$h_{out,6}$	4.88%	7.63	1.68
$h_{out,7}$	0	– 0.005	0.003
σ_1–σ_6	– 5.08	– 5.30	0.10
θ_1–θ_6	– 0.094	– 0.010	0.009
T_1–T_6	3.32	5.43	0.74
T_7	3.68	5.83	0.73

Table 5.15 Group 8 responses

Variables	Mean	Maximum deviation	Standard deviation
$h_{out,1}$–$h_{out,6}$	– 3.11%	– 5.26	1.10
$h_{out,7}$	0	0.005	0.002
σ_1–σ_6	– 5.07	– 5.26	0.09
θ_1–θ_6	– 0.010	– 0.11	0.004
T_1–T_6	3.32	5.43	0.74
T_7	3.68	5.83	0.73

As noted in the tables, the responses indicate reasonably good performance. This is seen in the almost negligible deviation in thicknesses in the outputs of the last stand. The small variations in the looper positions likewise imply little changes in the mass flows. This would indicate the likelihood of good performance throughout the mill, and likely an improvement over most conventional systems, with subsequent savings in costs and improvement in energy savings.

5.3 Lateral Control

Much of the control that is described in the literature is based on the longitudinal characteristics of the strip, with separate control for lateral characteristics. In general, it has been considered that to successfully realize performance in both control areas without significant interactions requires the longitudinal control to maintain a certain magnitude of the strip tension stress, and without undesirable perturbations. However, it is felt that more detail pertaining to areas of basic lateral control techniques would be useful for improvement of insight in these areas. Some of these methods are as follows and also provide examples for future updates to be considered for possible integration using the SDRE function for overall control of longitudinal and lateral characteristics.

In this description of lateral control functions it is assumed that two roll gap control cylinders are available on the mill stand, one on the operator side and one on the drive side[31] to realize a tilting of the rolls. In many instances (e.g., [26]) separate controllers are provided for each side with each independent of the main controller for strip thickness.

5.3.1 *Lateral Control, Example 1*

In this application [26] the main issue was the wide variety of products with different characteristics that must be successfully processed. In response to this, several groups of products were identified with a controller for each group. It was felt that a single linear quadratic (LQ) control would be inadequate for robust control for the entire set. Suitable models were developed, and in a typical instance seven groups of products were determined with a separately designed multivariable H^2 controller[32] for each group. The lateral position of the strip was determined by interstand cameras. The method developed was successfully applied on an actual mill with reasonably successful results. A MATLAB® toolbox was developed for additional use in other

[31] This is common practice in many mills. Some adjustments using the work roll jacks sometimes can be made, wherein there is a single cylinder for both drive side and operator side.

[32] The H^2 controller is based on a Hilbert space as defined in texts on robust control, e.g., [27].

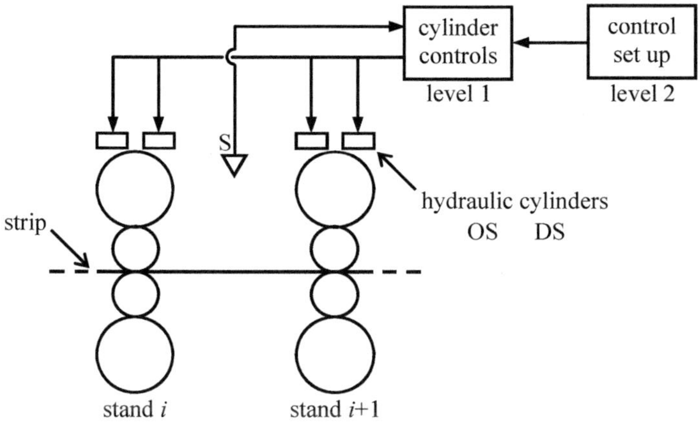

Fig. 5.45 Strip steering control. *S* strip lateral position sensor. *OS* hydraulic cylinder, operator side. *DS* hydraulic cylinder, drive side

Table 5.16 Strip deviation[a] from center

Control	Stand				
	1	2	3	4	5
H^2	12.5	9.0	7.4	7.3	2.8
[b]	14.0	16.0	19.0	20.5	8.3

[a] Standard deviation (mm) from off center values (100 strips by control, 200 by open loop)
[b] Open loop

mills. Figure 5.45 and Table 5.16[33] depict the actual installation and performance results.

The control seemed to work well in these situations. However, the control is based on mathematics that some users might find a bit more advanced, but which might be acceptable depending on the background of the users. In general this could be addressed on the basis of the particular application.

5.3.2 *Lateral Control, Example 2*[34]

Another method [28] involves the use of the automatic gauge control (AGC) function[35] combined with feed forward (FF) methods. In this technique the AGC, which

[33] Data in Table 5.16 are from [26].

[34] Copyright IEEE 2017, figures and tables as shown in this section. Portions of these are reprinted with cosmetic alterations, by permission, from [28]. Unless otherwise noted, data depicted are unchanged from [28].

[35] A description of a basic AGC function is in Chap. 4.

is common in many applications, is utilized to improve the performance and usage of the lateral control function. It is considered that this method has the potential for improved performance, does not require the use of somewhat more advanced mathematics, and provides some basic insight into certain lateral control issues. A brief description of this work and its results are as follows. A typical thickness description of the strip lateral profile is depicted in Fig. 5.46.

In the mill there are separate controllers for each cylinder, one for the operator side (OS) and one for the drive side (DS). In [28] three different approaches for the use of AGC to support and improve the lateral control are considered. In the first approach (AGC1) an average of the measured rolling forces at OS and DS, and an average of the measured cylinder positions at OS and DS, are used at both sides of the mill, so that the average value of the estimated thickness is used for both OS and DS. The result is that the control action for the AGC is the same for both OS and DS, which results in the avoidance of any tilting of the rolls due to AGC control.

In the second approach (AGC2) each of the controllers has a separate effect on its associated cylinder. Thus the responses of the AGC could be different for each side. This could be advantageous, especially where there could be asymmetric variations in the roll force. In these instances, such as variations in the strip hardness, this would be beneficial. However, there could be situations that are undesirable, such as excessive tilting of the rolls due to excessive uncertainties in the measurements that determine the length of the wedge, as well as certain asymmetric rolling situations. Further, any excessive tilting could result in an unsafe condition. Thus the possible use of the third approach appears to be more desirable.

In the third approach (AGC3) the addition of a feedforward (FF) technique was considered. A key assumption was the capability to determine the basic causes of asymmetric roll forces, and with some reasonably reliable means for the measurement of the causes. Some of the more probable situations and with possible measurements for FF, might be the deviation of the strip from center position, which can be

Fig. 5.46 Variables in a lateral cross section of strip, at the exit of a mill stand[36]

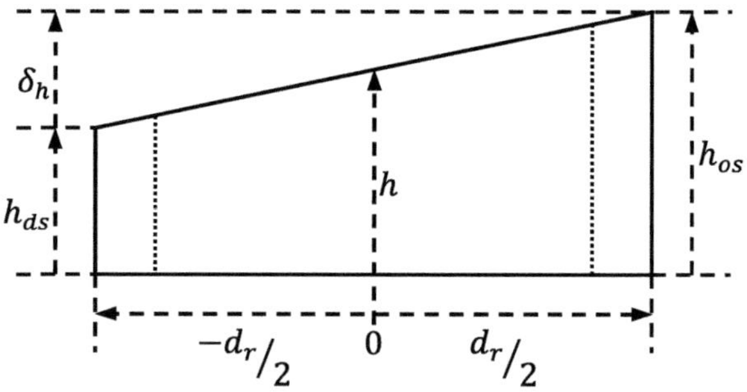

[36] In this evaluation the areas near the edges of the strip are not considered. This is because the multi-dimensional edge effects caused by deformation are unknown, and the effects of temperature drop near the edges can result in possible uncertainties in the modeling. For the purposes of this evaluation, the area in the center area of the strip was taken as the overall profile. Typical total uncertainties are about 15%, considering both sides [28].

determined by interstand cameras.[37] Also, undesirable profiles of strip temperature and strip thickness can be measured at the mill entry, with suitable tracking using observers (level 2) and with data from upstream stands, for FF use at downstream stands.

Additionally, the usage of FF can produce differences in the roll force references which can cause a trim in the hydraulic cylinder positions to improve the lateral profile, as depicted in Fig. 5.47 and noted in (5.39)–(5.42).

As depicted in Fig. 5.47 the FF trims (S_{FF}) cause adjustments in the DS and OS hydraulic cylinder position references (S_{c}). The adjustments in the cylinders by the FF and AGC trims at the DS and OS can be described as in (5.39) and (5.40):

$$S_{\mathrm{C,DS}} = S_{\mathrm{ref,DS}} + S_{\mathrm{AGC,DS}} + S_{\mathrm{FF,DS}}, \tag{5.39}$$

and

$$S_{\mathrm{C,OS}} = S_{\mathrm{ref,OS}} + S_{\mathrm{AGC,OS}} + S_{\mathrm{FF,OS}}, \tag{5.40}$$

where

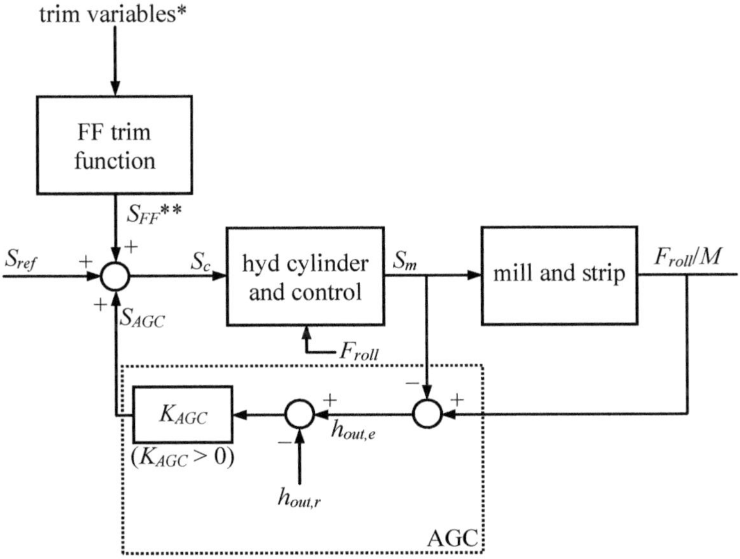

Fig. 5.47 Functional diagram of the basic AGC,[38] with feedforward trim function, typical for each of the OS and DS controllers. * Trim variables are the measured deviations in stand entry strip speed, thickness and temperature. ** S_{FF} is the FF trim on the cylinder position reference

[37] This measurement was found to be effective and useful in actual application as noted in Example 1.

[38] Additional detail is in Chap. 4, Fig. 4.8, and associated descriptions.

$$S_{\mathrm{FF,DS}} = \frac{\Delta F_{\mathrm{roll,FF,OS}} + \Delta F_{\mathrm{roll,FF,DS}}}{2M} - \frac{\Delta F_{\mathrm{roll,FF,OS}} - \Delta F_{\mathrm{roll,FF,DS}}}{2M_1}, \tag{5.41}$$

and

$$S_{\mathrm{FF,OS}} = \frac{\Delta F_{\mathrm{roll,FF,OS}} + \Delta F_{\mathrm{roll,FF,DS}}}{2M} + \frac{\Delta F_{\mathrm{roll,FF,OS}} - \Delta F_{\mathrm{roll,FF,DS}}}{2M_1}, \tag{5.42}$$

and where $\Delta F_{\mathrm{roll,FF}}$ is the deviation from the force reference, M is the mill modulus, and M_1 is a separate mill modulus for the difference of measured forces at the DS and OS.[39] Thus the FF function can be used with both of the previously described first and second approaches. The FF function handles the asymmetries, and thus it can be used with the first approach. The remaining errors in thickness, such as plant-model mismatch, measurement disturbances, and other sources of strip inhomogeneities that can't be measured, are handled by the AGC. Thus the use of the first approach with FF is preferred over the second approach.

What follows is an example of how the FF technique can handle a lateral off center position of the strip. In this instance the strip position is taken to be off center, wherein the lateral position of the center of the strip is C_{m} (measured and positive in the direction of the OS), d_{cr} is the distance between the hydraulic cylinders, and thus, per Fig. 5.46, the strip edges are at $(- d_{\mathrm{r}}/2 + C_{\mathrm{m}})$, and $(d_{\mathrm{r}}/2 + C_{\mathrm{m}})$, and $F_{\mathrm{roll,DS}}$ and $F_{\mathrm{roll,OS}}$ are forces at the operating point. Then, taking the force across the rolls as being uniformly distributed, the additional roll forces are approximately

$$\Delta F_{\mathrm{roll,OS}} = \frac{C_{\mathrm{m}}}{d_{\mathrm{cr}}}\left(F_{\mathrm{roll,DS}} + F_{\mathrm{roll,OS}}\right), \tag{5.43}$$

$$\Delta F_{\mathrm{roll,DS}} = -\frac{C_{\mathrm{m}}}{d_{\mathrm{cr}}}\left(F_{\mathrm{roll,DS}} + F_{\mathrm{roll,OS}}\right). \tag{5.44}$$

Thus, with the appropriate control, the FF method can be used to handle the off center position, which also can be useful for handling incoming asymmetrical characteristics such as temperature or thickness.[40]

Several situations were addressed in the simulations to get a better feel for the response to various events. These are the strip off center, errors in the measurement of the roll force, maintaining a desired wedge profile, and asymmetry in the entry temperature and thickness.

In the case of the strip off center, the three control approaches noted previously as AGC1, AGC2, and AGC3 were considered. For this simulation it was assumed that the strip was off center and moving toward the drive side under AGC1 control. As

[39] During simulations it was determined that in the negative relationship of the roll force trims, a separate mill modulus was needed. This apparently occurred as the rolls were less stiff, and thus were more sensitive to bending and flattening. This also would be more likely in the measurements of a wedge-shaped strip between OS and DS, and in the negative relationship of the trims. If the rolls were stiffer, the deviation would be realized more by the mill housing than the roll stack.

[40] More detail and data are available in [28].

Table 5.17 Strip deviation from center

Control	Wedge length[a]
AGC1	– 2.75 mm
AGC2	– 0.027
AGC3	0

[a] Approximate change in length of wedge (δ_{h}) after 3 s during movement toward DS from initial position; positive movement is toward OS; data are derived from [28]

the strip moves, the roll force $F_{\text{roll,DS}}$ on the drive side increases, and $F_{\text{roll,OS}}$ on the operator side decreases, which is expected since ΔF_{roll} is close to zero, according to AGC1 characteristics. This results in a strip wedge (example δ_{h} in Fig. 5.46) that continually gets larger, since there are no separate control actions from AGC1. In the case of AGC2 wherein separate control actions are used, the rolls are allowed to tilt and the effect is somewhat improved. In the case of AGC3, there is much less deviation due to the FF action, as an example in Table 5.17.

In the case of errors in measurement of the roll force, it was assumed that there was an error in the measurement of the roll force at the DS. A possible cause could be asymmetric friction in the mill stand. For simplicity of understanding, it was assumed that the strip was centered in the roll gap. In the cases of the control by AGC1 or AGC3, due to the error in the force measurement, there could be a difference in the positions of the DS cylinder and the OS cylinder, which was caused by the difference in the actual forces. Thus the effects of each of the three control efforts would result in a change in the average strip thickness. However, in the case of control by AGC2, the controller would additionally tilt the roll stack, which would result in an additional error in the length of the wedge, and thus be less desirable.

In the case of a desired wedge profile, the shape of the wedge is established by the desired value of δ_{h} as in Fig. 5.46. To realize this, the average value of the desired exit thickness is held constant. This involves modifications of the present values in (5.39) and (5.40) as follows so that the desired values of the thickness references are adjusted as in (5.45) and (5.46), where

$$S_{\text{C,DS}} = S_{\text{ref,DS}} + S_{\text{AGC,DS}} + S_{\text{FF,DS}} + S_{\text{W}}, \tag{5.45}$$

$$S_{\text{C,OS}} = S_{\text{ref,OS}} + S_{\text{AGC,OS}} + S_{\text{FF,OS}} + S_{\text{W}}, \tag{5.46}$$

and

$$S_{\text{W}} = \frac{\delta_{\text{h}}^{d}}{2}\frac{d_{\text{dr}}}{d_{\text{r}}}, \tag{5.47}$$

where S_{W} is a change to the reference to account for the wedge, δ_{h}^{d} is the desired wedge dimension, d_{dr} is the distance between the cylinders, and d_{r} is the strip width (Fig. 5.46). It can be noted that this change in the wedge dimension can be realized along with the compensation for concurrent asymmetries, which are linearizations around the associated operating point.

In the case of entry asymmetric temperature of the strip, for the simulation the average temperature across the width of the strip was assumed to remain constant, but with deviations in the temperatures at the OS and the DS, so that the higher value of temperature would result in a lower yield stress. Thus, if the positions of the cylinders remained unchanged, the exit strip thickness at the OS is lower than at the DS, and thus a lower force at the OS will result in a lower mill stretch, with a smaller estimated error in the strip wedge. However, the expected effect on the wedge dimension is very small, so that the control by AGC2 is less effective, so that the feedforward effect of AGC3 would be the preferred method.

The previous results can be applied to other asymmetrical characteristics of the incoming strip, such as those due to lateral deviations in thickness, assuming that the characteristics are known or can be reasonably estimated. Additionally, it shows the potential advantages of feedforward methods when using applicable AGC methods. It also can be noted that the previously described method using FF techniques can reduce the effects of asymmetric strip tension which has the potential for undesirable interactions with the downstream looper.

5.3.3 Lateral Control, Example 3

This example [29] expands on the basics presented in Example 2 to include a work roll bending system to consider the effects of the deflection of the work rolls and the control of the strip crown. Additionally, an optimization function is presented for an improvement in feedforward (FF) control.

In this example two methods of control are addressed as to their capabilities to improve performance and with a significant reduction in control error. The methods considered are single-input–single-output (SISO) and multi-input–multi-output (MIMO) strategies for implementation of FF techniques. A brief description of this work and its results are as follows.

In the case of the SISO system as in Example 2, linearized models for the roll force and the yield stress of material are used. The variables of interest are the temperature and thickness of the strip at the entry of the first mill stand, plus the strip speed. The strip temperature and thickness are obtained by measurements at the strip entry, while the strip speed is obtained using the work roll characteristics and the characteristics of its drive. These variables are used in the FF function (Fig. 5.47).

At an operating point the deviation in the roll force can be expressed using deviations in strip thickness, temperature, and speed:

$$\Delta F_{\text{roll,FF}} = \Delta h_{\text{ent}} \frac{\partial F_{\text{roll}}}{\partial h_{\text{ent}}} + \Delta T_{\text{ent}} \frac{\partial F_{\text{roll}}}{\partial T_{\text{ent}}} + \Delta v_{\text{ent}} \frac{\partial F_{\text{roll}}}{\partial v_{\text{ent}}}, \tag{5.48}$$

Fig. 5.48 Coordinate axes at the upper backup roll at the exit of a mill stand[41]

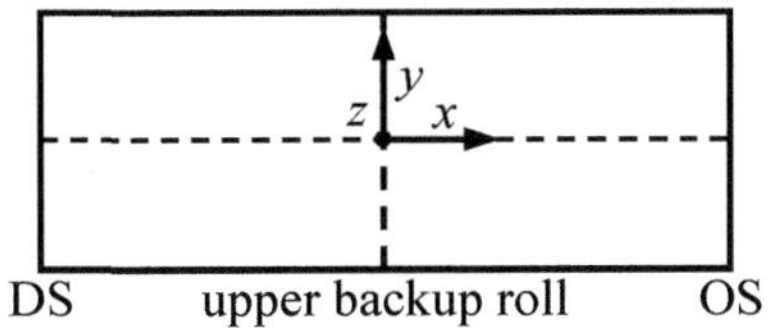

where $\Delta F_{\text{roll,FF}}$ is the (expected) deviation in the roll force from the operating point as used in the FF function; h_{ent}, T_{ent}, and v_{ent} are variables representing the strip thickness, strip temperature, and strip speed which are measured values at the lateral center of the strip. The deviation $\Delta F_{\text{roll,FF}}$ causes a change in the mill stand which is compensated by a change in the cylinder position via the FF trim S_{FF} (Fig. 5.47):

$$\Delta S_{\text{FF}} = \frac{1}{2}\frac{\Delta F_{\text{roll,FF}}}{M} - \Delta h^{d}_{\text{ext}}, \tag{5.49}$$

where $\Delta h^{d}_{\text{ext}}$ is the anticipated change in the exit thickness and M is the mill modulus. This compensation is applied to both the OS and the DS, so that the FF compensation is symmetric.

An advantage of the SISO system is its simplicity, which makes it reasonably easy to understand and use, and the measurements needed (Fig. 5.47) are available. However, the basis of the system is a linearized model of the roll force, which significantly limits the useful range of operation to a smaller area around the operating point. And also of importance are the inability to handle variations in the lateral thickness profile, the various conditions wherein the strip is not in the lateral center of the roll gap, the situations wherein asymmetric control of the OS and DS control is required, and also the use of the work roll bending jacks. Thus, while some of these requirements are addressed previously in Example 2 as a basis, a more advanced approach is needed to provide a more fully useful and effective method, such as an optimized MIMO FF control strategy as described in the following example.

In this example of control a strategy [29] is developed that compensates for upstream disturbances using FF techniques. The inputs to the optimized controller are the positions of the OS and DS hydraulic cylinders and the forces of the OS and DS work roll bending jacks, which are elements of an input vector u, where

$$u = \left[S^{\text{DS}}_{\text{m}}, S^{\text{OS}}_{\text{m}}, F^{\text{DS}}_{\text{wrj}}, F^{\text{OS}}_{\text{wrj}}\right]', \tag{5.50}$$

where S_{m} and F_{wrj} are the cylinder positions and the work roll bending forces developed by the bending jacks, and (′) indicates the matrix transpose. The disturbances being considered are the entry thickness profile $h_{\text{ent}}(x, z)$ and the entry temperature profile $T_{\text{ent}}(x, z)$ of the strip, with the (x, z) axes as depicted in Fig. 5.48 at the upper backup roll, which are mapped to the strip cross section in the roll gap. The objective

[41] Copyright 2018 Elsevier. Portions of Fig. 5.48 are reprinted with cosmetic alterations, with permission, from [29].

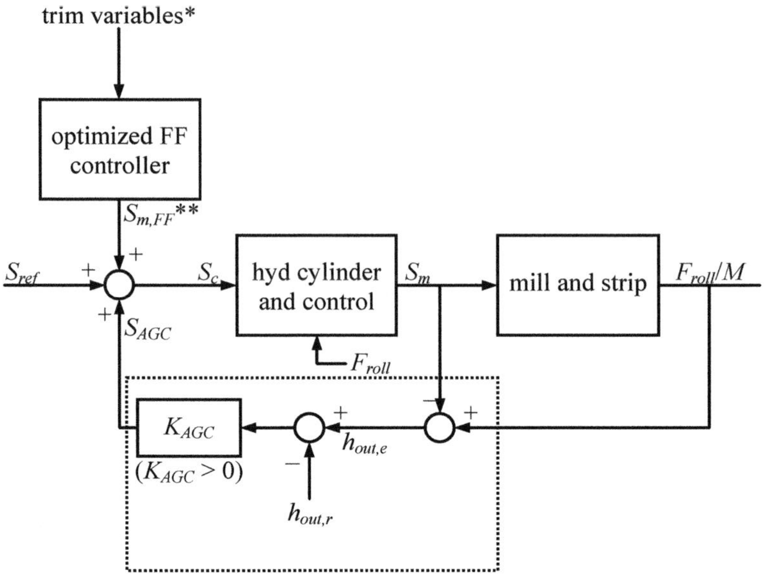

Fig. 5.49 Functional diagram of the basic AGC, with optimized controller, typical for each of the OS and DS controllers.[42] * Trim variables are the measured deviations in stand entry strip thickness and temperature. ** $S_{\mathrm{m,FF}}$ is the FF trim of the optimized controller on the cylinder position reference S_{ref}

of this work is to obtain a cross sectional profile of the strip with very low dimensional uncertainties.

To reduce the dimensional uncertainties in the strip significantly an optimization function (Fig. 5.49) was applied to minimize the error in the profile of the lateral thicknesses. A quadratic cost function based on the deviation of the actual strip profile from the desired strip profile was developed [29] as in (5.51), for example as follows:

$$h_{\mathrm{err}} = \left(\frac{1}{d_{\mathrm{r}} - 2\overline{d}} \int_{-\frac{d_{\mathrm{r}}}{2}+d}^{\frac{d_{\mathrm{r}}}{2}-d} \left(h_{\mathrm{ext}}(x) - h_{\mathrm{ext}}^{\mathrm{d}}(x)\right)^2 \mathrm{d}x \right)^{\frac{1}{2}}, \tag{5.51}$$

where d_{r} is the strip width as in Fig. 5.46, and $\overline{d}$ is a positive adjustable constant that can be set to correspond to the particular strip width being processed, $h_{\mathrm{ext}}(x)$ and $\Delta h_{\mathrm{ext}}^{\mathrm{d}}$ represent the actual and desired thicknesses, and h_{err} is the thickness error that is to be minimized. The minimization is with respect to u (5.50) and with the requirements:

[42] Copyright 2018 Elsevier. Portions of Fig. 5.49 are reprinted with cosmetic alterations, with permission, from [29].

$$h_{\text{ext}}(x) = f_{\text{sm}}\left(S_{\text{m}}^{\text{DS}}, S_{\text{m}}^{\text{OS}}, \dot{q}_{\text{roll}}, F_{\text{wrj}}^{\text{DS}}, F_{\text{wrj}}^{\text{OS}}, s_{\text{wrp}}, v_{\text{roll}}\right), \tag{5.52}$$

$$f_{\text{roll}}\left(q_{\text{roll}}, h_{\text{ent}}, h_{\text{ext}}, \sigma_{\text{ent}}, \sigma_{\text{ext}}, v_{\text{roll}}, k_{\text{yfm}}\right) = 0, \tag{5.53}$$

with constraints:

$$0 \ll S_{\text{m}}^{\text{n}} \ll S_{\text{m}}^{\text{max}} \text{ and } 0 \ll F_{\text{wrj}}^{\text{n}} \ll F_{\text{wrj}}^{\text{max}}, \tag{5.54}$$

and where $h_{\text{ext}}(x)$ is the strip exit thickness at x; f_{sm} is the mill stand model; S_{m} is the cylinder position; DS and OS are drive and operator sides; $\dot{q}_{\text{roll}}$ is the time derivative of the local roll force; F_{wrj} is the work roll bending force; s_{wrp} is the work roll shifting position; v_{roll} is the rolling velocity; f_{roll} is the roll gap model; h_{ent}, h_{ext}, are entry and exit thicknesses; σ_{ent}, σ_{ext}, are strip tensions on entry and exit sides; k_{yfm} is the yield stress; *max* is a maximum; and *n* applies as DS and OS.

The solution to this quadratic problem is reasonably straightforward using well-known methods. In the case of actual application to the tandem hot mill, in general a solution must be repeatedly solved during each scan of the controller for equipment used in actual implementations. Typically a scan is performed in about every 15–20 ms for tandem hot mills, so that the solution of the quadratic problem in each scan was beyond the capability of certain presently existing equipment. Thus to conform to this requirement, certain modifications to the solution were made, which included approximations, such as linearization in certain areas. The details are a bit extensive and required additional application of somewhat more advanced methods, such as those in [29]. Thus, in the case of similar applications to tandem hot rolling for use with present equipment, this requirement should be considered during the controller design.[43] A diagram of a functional application of the MMO FF controller is depicted in Fig. 5.49.

Several simulations were performed. In this instance the performance of the controller was evaluated at the entry mill stand, with the objective of evaluating the controller to produce ideal thicknesses of the strip in lateral and longitudinal directions in under somewhat more severe conditions.

In certain initial simulations which follow, the differences between the linearized mill stand model and the complete model were addressed. The results appeared to be acceptable with only minor variations in performance. However, also it can be noted that in these cases using the overall optimal FF control strategy, the measurements of roll force and cylinder position were not used, and the entry strip thickness and temperature were taken to have very low uncertainties to avoid differences between the plant model and the actual plant. Thus this required the AGC function to handle only any remaining uncertainties (of lesser magnitude), which appeared to perform reasonably well and nearly similar to that of the SISO FF controller. More specific detail is described and addressed in what follows and in [29].

[43] In the case of using the SDRE method for control in the longitudinal direction, this requirement is successfully implemented as described in Sect. 5.2.4. Investigation of possible similar usage of SDRE techniques for the inclusion of lateral control would be worthy of future consideration.

For the basis of the simulations, the characteristics of a strip at the mill entry were based on measurements taken from an actual mill. The temperature measurements were made by thermo-graphic cameras and pyrometers with the strip thicknesses using an infrared laser and X-ray unit. In addition, to better evaluate controller performance under asymmetric rolling conditions, the measured temperature gradient laterally was increased by adding an increase to the measured temperature at the DS, while similarly and concurrently decreasing the same measured temperature at the OS. The intent was to simulate the actual effects during the processing of slabs in the reheating furnace, wherein a significant temperature gradient can occur by a slab being near a furnace wall or door. The overall intent was to test the controller's performance under more severe situations. Some characteristics of the incoming strip are as follows.[44]

Simulation 1 was performed to evaluate the uncertainties in the use of a linearized mill stand model versus the full model in the presence of the strip characteristics as described in Table 5.18. Trims to the references were included for the cylinder positions on the OS and DS. The references for the work roll bending jacks were without trims for the SISO FF controller. At temperatures below the 300 m length and at colder areas of the strip, an increase in the trims of the cylinder positions and in the bending forces of the work rolls was noted, with lesser asymmetry in the trims. As would be expected as the temperature asymmetry was increased at lengths greater than 300 m, the asymmetry was increased between the cylinder positions on the OS and DS, and between the work roll bending jacks at OS and DS. Also, it could be noted that there was very little difference between the linearized model and the actual model. Overall the system performance was reasonable for this particular strip and within the description of Table 5.18.

Simulation 2 was performed to evaluate the differences between AGC control with optimal MIMO FF, AGC with SISO FF, and AGC alone. In each instance the responses are based on the process and control descriptions of Table 5.18, Figs. 5.47, and 5.49. Further, it is assumed that the plant model ideally matches the actual plant.[45] The results of this simulation are summarized in what follows.

The deviations in the AGC feedback were compared with trims from the MIMO FF or SISO FF, as depicted in Fig. 5.49. In these instances, the FF trims were dominant with respect to the AGC feedback. This might be expected since the AGC feedback has an associated time lag due to control action when compared to the FF, and the linearized model closely represented the actual plant.

The corresponding output thicknesses responded closely to the responses of the FF trims and the AGC feedback, with similar deviations. For example, the maximum deviation in the simulated output thicknesses for the AGC was about 0.05 mm and was increasing after 600 m length of strip, with a desired steady-state thickness of 21.5 mm. The deviation for the SISO FF was less than 0.02 mm immediately after

---

[44] Data as available were derived from [29], and are approximate. Additionally, the temperature difference was deliberately increased at a length of 300 m to examine controller performance under more severe conditions.

[45] In these examples the effects of the plant-to-model mismatch by linearization were less significant as the deviations from the actual plant were only minor.

Table 5.18 Characteristics at Stand 1 input

Strip length, speed, temperature	Length 600 m, width 1.0 m, strip is in stand after threading
	Speed increased (approximately linearly) from 1.05 to 1.44 m/s
	Average temperature about 1100 °C at strip center
	Temperature increase from OS to DS, about 50 °C, length < 300 m
	Temperature increase from OS to DS, about 85 °C, length > 300 m
	Temperature deviations caused by skid marks, about 10 °C to about 25 °C, length < 300 m
	Temperature deviations caused by skid marks, undetermined for length > 300 m
	Distance between skid marks, about 110–150 m, length < 300 m
Strip thickness	Average thickness about 37.0 mm at strip center
	Increase in thickness from OS to DS, about 1.5 mm, length < 300 m
	Increase in thickness from OS to DS, about 1.5 mm, length > 300 m
	Thickness deviations caused by skid marks, from near zero up to about 0.7 mm
	Skid marks are at positions similar to temperature skid marks

the initial strip length and negligible after about 200 m of strip length. The deviation for the MIMO FF was negligible from the initial strip length and remained so for the entire strip length. In the case of the skid marks, they were negligible in the case of MIMO FF, were very slightly noticed in the case of the SISO FF, and still remaining in the case of the AGC alone.

Additional simulation was done to evaluate the capability of the three control schemes to retain a desired thickness profile over the width of the strip and also over its entire length. A measure of the capability of a control method can be somewhat realized by evaluating the error function (5.51) and also its change in width over the strip length. The desired profile was best achieved when the MIMO FF method was used as in Fig. 5.49. This is because in this method, asymmetric control of the OS and DS functions is possible to realize the required thickness profile, which is not so in the case of the previously noted basic SISO FF method with a trim on the cylinder position, as the trims are symmetric, i.e., the trim on the OS is the same as the trim on the DS, except in the opposite direction. Additionally, and similarly in the case of the AGC alone method, the controller was also simple proportional, and thus has an inherent offset between the reference and the feedback.

Some examples are follows. In the cases of the error using the optimal MIMO FF, the overall maximum error was only slightly above zero.[46] In the case of the SISO FF, the error varied between 20 μm initially and decreased to about 5 μm at about 300 m, and then was steadily increasing as the temperature was raised during the strip length beyond 300 m. In the case of the AGC alone, the error varied between 5 and 35 μm before the strip length of 300 m, and varied between 5 μm and above 45 μm beyond the strip length of 300 m, mostly due to the effect of skid marks.

It can be noted that in this case, where there is an asymmetric temperature condition across the strip, an undesirable wedge shape can be produced. This can occur wherein symmetric control is used, e.g., SISO FF or AGC alone, which produces symmetric control inputs. In this case the mean strip thickness remains equal to the desired thickness, but the profile across the strip is no longer the same, and a wedge[47] shape of the strip is produced. Thus an undesirable camber shape can result, with a lateral strip movement, which can increase over the length of the strip. As an example, in the case of the strip characteristics described in Table 5.18, with an the asymmetric temperature characteristic in the strip length from 300 to 625 m, the use of an AGC alone control results in a continually increasing wedge in the strip from 0.05 mm to about 0.12 mm. This would be avoided with the MIMO FF controller which produces asymmetric trimming functions. An example as adapted from [29] is provided in Fig. 5.50.

As can be seen in Fig. 5.50, the MIMO FF w/AGC handles the wedge quite well, with the SISO FF w/AGC being close to the desired profile, with the AGC alone having a larger deviation.

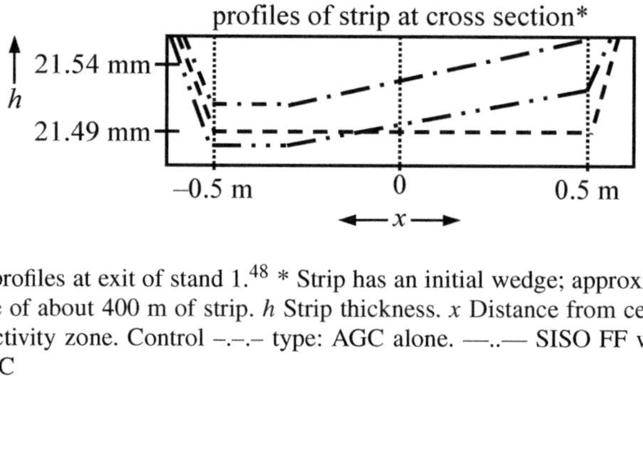

Fig. 5.50 Strip profiles at exit of stand 1.[48] * Strip has an initial wedge; approximate profiles are just after passage of about 400 m of strip. *h* Strip thickness. *x* Distance from center line. [− 0.5, 0.5 m] control activity zone. Control –.–.– type: AGC alone. —..— SISO FF w/AGC. – – – – . MIMO FF w/AGC

[46] This offset occurred because only a few samples were taken to control the load over the entire length of the roll.

[47] A wedge is the difference in thickness between the DS and the OS of the strip, e.g., Fig. 5.46.

[48] Copyright 2018 Elsevier. Portions of Fig. 5.50 are reprinted with cosmetic alterations, with permission, from [29].

5.3.4 Additional Real Plant Application

In this section an overview, based on [29], is presented. The measured results of the simple SISO FF strategy, as described more fully in [28], as applied at the first mill stand were considered.[49] In this particular case the speed of the strip was not considered at the request of the plant operator. This was because only the sensitivities for strip thickness and temperature in (5.48) were available and well-known by the plant operator for several materials, so that (5.48) becomes (5.55):

$$\Delta F_{\mathrm{roll,FF}} = \Delta h_{\mathrm{ent}} \frac{\partial F_{\mathrm{roll}}}{\partial h_{\mathrm{ent}}} + \Delta T_{\mathrm{ent}} \frac{\partial F_{\mathrm{roll}}}{\partial T_{\mathrm{ent}}}. \tag{5.55}$$

Thus the expected variations in the roll force can be somewhat reasonably estimated by measuring deviations from the operating point by measuring the deviations in the entry temperature and the entry thickness. In this case the entry thickness was determined by measurement at the output of the roughing mill while the entry temperature was measured by two pyrometers, one at the upper and one at the lower surfaces of the strip, with an average taken as the strip temperature. The lateral characteristics of the strip thus were unavailable at this point. However, it was assumed that the temperatures and strip thicknesses were about the same at the mill entry as at the point of measurement, and that there was sufficient time for computation of a FF trim, e.g., $S_{\mathrm{m,FF}}$ in Fig. 5.49.

To verify the simulations of FF, two strips with very similar characteristics were processed consecutively. In both cases the FF trims on the cylinder position reference ($S_{\mathrm{m,FF}}$ and S_{ref} in Fig. 5.49) were determined by calculation using the simulated FF control system. In one case the measurement of the actual FF trim was taken for one of the strips and compared to the results of the calculations. The results during the processing of data from both strips were compared. The results showed similar responses during the same length (about 500 m) of strip, which confirmed in this case the usefulness of the FF method and the reasonable validity of its simulations.

Additional simulations were performed with two strips of very similar characteristics. The first strip was processed with the FF trim and the AGC feedback. The second strip was processed with AGC feedback alone. In the case of the first strip, most of the deviations in the FF trims were sufficient to handle well the deviations in the skid marks, with only minor deviations in the AGC feedback, and nearly negligible skid mark deviations, which thus confirmed the usefulness of the FF.

The simulations were continued with many additional strips. To evaluate the usefulness of the FF method, a standard deviation $\sigma_{\mathrm{sd,h}}$ over the strip length was computed as in [29],

[49] The application was the finishing mill at the Voestalpine site in Linz, Austria. More detail is available in [29].

$$\sigma_{\mathrm{sd,h}} = \sqrt{\left(\frac{1}{n_z - 1}\sum_{l=1}^{n_z}\left(\Delta h_{\mathrm{ext}}(z_l) - p_{\mathrm{sd,h}}(z_l)\right)^2\right)}, \tag{5.56}$$

where z is the longitudinal length along the strip length, $h_{\mathrm{ext}}(z_l)$ is the expected strip thickness at the output of the first mill stand at the point z_l of the strip, $p_{\mathrm{sd,h}}(z_l)$ is a first-order polynomial that masks out the increase of the strip thickness due to the increase in the strip speed,[50] and n_z is the number of points along the strip length. The standard deviation was computed for two groups of the additional strips. The first group of 2250 strips was processed using the AGC with SISO FF trim, while the second group of 3200 strips was processed using the AGC alone. The frequency distribution was determined for both groups. The results were that the mean value for the FF control was lower than that of the AGC alone, i.e., a mean value of 0.077% versus 0.118%,[51] which implies that large deviations from desired thickness are significantly reduced by using the SISO FF method. Thus there are reasonable possibilities of even further improvement with MIMO FF AGC systems, or other advanced MIMO methods such as potential application using SDRE techniques, with a fully integrated SDRE system for control of the entire mill.

5.3.5 *Comment*

As might be expected, there are other methods for lateral control, some of which are referenced in [28, 29], plus some that are a bit more advanced. The methods as described previously, and as described more fully in [28, 29], are included to show examples of some of the basic situations that must be considered in this area, and with useful advanced methods. It is intended that what is presented would provide some basic insight into control of this area, which might be expanded upon to realize further useful advancements.

5.4 Model Predictive Control

The function of model predictive control is to determine control inputs by using an optimization criterion that is developed based on a prediction horizon. A system model is developed in state-space form and is usually linear although some nonlinear models have been successfully applied. In most instances a discrete time model is used.

---

[50] The sensitivities of strip speed could not be included as the plant operator had none available for the material being processed, as noted in (5.55).

[51] Available data were derived from [29].

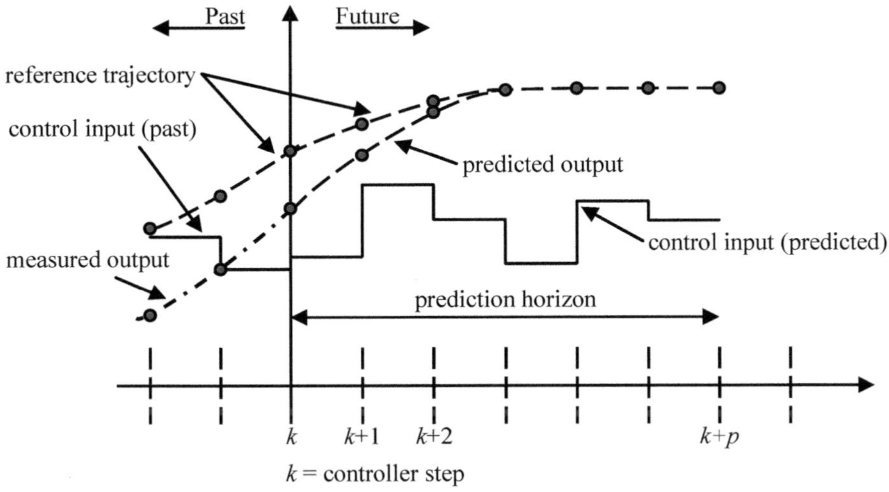

Fig. 5.51 Model predictive control functional diagram

The main idea is to predict the effects of future inputs on the states of the system and its outputs. This is accomplished by the formation of an objective function using the process model. At each discrete step of the controller, a move of the controller is computed for each step in a prediction horizon. This is done in a manner to minimize the objective function, and to estimate future values of the states and outputs, while the model (in the form of the state equation) is used as a constraint during the move. However, while inputs are computed for each step, only the first step of inputs is actually used. The prediction horizon is then moved forward one step and the procedure is repeated using measurements from the first movement of the controller. Figure 5.51, [30, 31] provide some further detail.

Model predictive control has been used successfully in other applications, particularly in the chemical processing industries, and it appears to offer some advantages, such as possible improvements for nonlinear control. However, it is not clear at present that it has been shown to fully realize a significant improvement beyond some of its presently proposed methods for nonlinear control of the entire tandem hot mill. This is because the controller should successfully handle a large nonlinear system and have a sufficiently fast update time of about 15–20 ms which is typical for other existing systems that presently control the mill. Additionally, the system should be able to perform well over the entire operating range which includes a wide variety of various products, during normal and abnormal situations, and in the presence of significant disturbances and uncertainties. Further, the controller should be user friendly to commissioning personnel who have lesser backgrounds in advanced control theory.

It can be noted that some initial work in this area has been done for thickness and looper control (e.g., [32]), and model predictive control appears to have the potential for useful application to larger nonlinear systems such as the tandem hot mill. However, further additional efforts in the development of this method for control of the mill over the entire operating range of the mill could require fully addressing what appear to be needed advancements, and with any needed improvements in available hardware and software.

5.5 Sliding-Mode Control

Sliding-mode control is a method based on a switching function that switches a group of control laws. The switching, as determined by the switching function, is based on the current behavior of the system, so that the system's control law is selected in accordance with some predefined rules that depend on the system's state.

To accomplish this, a manifold with dimensions lower than those of the original system is determined such that it can be made stable (and positive invariant) by a suitable control input. Next, a control law is chosen that forces the states of the system that are not on the manifold to move to the manifold within a finite time. The control is switched between an upper bound and a lower bound, wherein the logic for switching is a function of the distance from the manifold. Now because the manifold is invariant, the trajectory remains on the manifold, and there are no longer uncertainties in the inputs. This manifold is denoted as the sliding-mode.

Unfortunately, due to the process characteristics, such as time delays and hysteresis, the states usually never remain on the manifold, but repeatedly cross it. This undesirable effect is known as chattering. To try to eliminate or reduce the effects of chattering, a boundary layer is established around the manifold so that the states remain within the boundary layer. However, to fully accomplish this, the boundary layer can become quite wide, such that robustness to uncertainties is significantly reduced, and in some instances nearly to the point of being useless.

In smaller industrial applications which require robustness to uncertainties in modeling, and with disturbances, sliding-mode control has been quite effective. Similarly, in the case of tandem hot mill rolling applications, in specific areas it has been useful, or appears to have useful potential (e.g., [33, 34]). However, in larger applications for the entire mill it is much less effective. This is because a much more complex application is needed due to many more parameters and undefined uncertainties and disturbances, many of which are also interactive. This is especially so since any chattering is unacceptable as it could result in excessive wear on mechanical components such as hydraulics, and thus it becomes extremely difficult, if even possible, to operate without an effective boundary layer.

It is recognized that recently some work has been done that claims to fully eliminate chattering (e.g., [35, 36]). However, to be applicable to the entire area of tandem hot rolling, it appears that additional work needs to be done to be fully useful for larger applications. More detail on sliding-mode control is available in [37, 38].

5.6 *H*-Infinity Loop Shaping

H-infinity (H_∞) control is a method wherein uncertainties and disturbances in linear time invariant multivariable systems are addressed in a systematic manner. In this method, a space (denoted as the Hardy space) consists of the entire group of matrices that represent the stable and proper transfer functions of a particular system. In this system the infinity norm of a particular matrix is minimized, and associated procedures are applied to develop weighting functions; then, using optimization techniques, a stabilizing controller is determined.[52]

The H_∞ loop shaping method is similar to the classical loop shaping technique. In the classical technique, methods of loop shaping are applied to realize a controller that will guarantee stability and performance in the presence of disturbances, and with uncertainties such as those resulting from model mismatch. In this method any differences resulting from model mismatch are described in the frequency domain using appropriate frequency dependent bounding functions. The control designer determines the desired performance by developing weighting functions that are added to the plant model to describe the desired performance, so that in the frequency domain the magnitude and phase of the transfer function of the augmented plant model always remain within the established bounds at all frequencies.

In the H_∞ loop shaping method for control of linear time invariant multi-input–multi-output (MIMO) systems a similar technique is used, except using open loop matrix singular values. In this case weighting matrices are added to the plant-model matrix to approximate the desired characteristics of the modified plant. Then, using H_∞ optimization techniques an interim controller is developed to stabilize the modified plant. The final controller for the original plant is developed by using the weighting matrices and the interim controller.[53]

This method is often used for control of linear time invariant MIMO systems, for example in the design of certain aircraft control systems; and it might initially appear to be useful for control of the tandem hot mill.[54] However, establishing the boundaries of the singular values, and the determination of the weights are some of the complex tasks which require an experienced and intuitive designer. Moreover, the tandem hot mill system is large, highly nonlinear, and with a large range of disturbances and uncertainties that can interact with each other, which makes the method even less desirable in most instances. Thus, considerable improvement appears to be required for successful application to the tandem hot mill and would need to be addressed in the future investigations.

---

[52] The definition of the infinity norm, determination of weighting matrices, optimization methods, and more detail is available in texts on robust control (e.g., [27, 39]).

[53] Much more detail is available in [40] and the references cited therein.

[54] An example of a proposed application to a tandem cold mill is described in [41].

5.7 Concluding Comments

Four major methods of what might be considered as advanced control were addressed. Of these four, it was felt that the SDRE technique has the most likelihood of providing significant improvement in performance, while yet being reasonably user friendly to design and commissioning personnel. This is due to its general simplicity of design and implementation, and not requiring major advanced mathematical details. Of course, not every possible method can be addressed, but of these four the SDRE appears to have the most advantages for a larger nonlinear application with major uncertainties and disturbances, the need for a fast update time in its control, a wide variation in the material being processed, plus continual customer requests for increases in stringent requirements in the final product. Moreover, the usefulness of the SDRE technique in other practical applications has recently become very widely accepted in a variety of other areas as noted in the cited references. Thus it is felt that the use of the SDRE method, in addition to its present advantages, has a significant potential for further improvements in the control of tandem hot metal strip rolling applications.

Appendix

Optimality of Certain Applications

As noted in Sect. 5.2.2, the necessary conditions for optimality of an SDRE controller are fairly complex, but sometimes can be realized in a local area. The following provides some brief detail.[55]

Considering the state-dependent control problem:

$$\dot{x} = A(x)x + Bu, \quad t \in [0, t_f) \quad x(0) = x_0, \tag{5.57}$$

and cost function:

$$J = \frac{1}{2}\int_{0}^{t_f} \left(x'Q(x)x + u'R(x)u\right)\mathrm{d}t. \tag{5.58}$$

The Hamiltonian approach to derive the necessary conditions for the optimal control are summarized according to the following steps:

$$H(x, u, \lambda) = \frac{1}{2}x'Q(x)x + \frac{1}{2}u'R(x)u + \lambda'A(x)x + \lambda'Bu, \tag{5.59}$$

---

[55] The symbol ($\dot{z}$) indicates the time derivative of the variable z; the symbol ($'$) indicates the matrix transpose, the symbol (∇) indicates the gradient.

$$\dot{x} = \nabla_\lambda H = A(x)x + Bu, \quad x(0) = x_0, \tag{5.60}$$

$$\begin{aligned}\dot{\lambda} = &- \nabla_x H = - Q(x)x - \frac{1}{2}x'\frac{\partial Q(x)}{\partial x}x - \frac{1}{2}u'\frac{\partial R(x)}{\partial x}u \\ &- A'(x)\lambda - \lambda'\frac{\partial A(x)}{\partial x}x, \quad \lambda(t_f) = 0,\end{aligned} \tag{5.61}$$

$$0 = \nabla_u H = R(x)u + B'\lambda. \tag{5.62}$$

The notation $x'\dfrac{\partial M(x)}{\partial x}x$ (as in 5.61) for an arbitrary $n \times n$ matrix $M(x)$ refers to the following $n \times 1$ vector

$$x'\frac{\partial M(x)}{\partial x}x = \begin{bmatrix} x'\frac{\partial M(x)}{\partial x_1}x \\ x'\frac{\partial M(x)}{\partial x_2}x \\ \vdots \\ x'\frac{\partial M(x)}{\partial x_n}x \end{bmatrix}, \tag{5.63}$$

where the matrix $\dfrac{\partial M(x)}{\partial x_i}$ is an $n \times n$ matrix whose ijth entry is $\dfrac{\partial M_{ij}(x)}{\partial x_i}$.

From (5.62), $u = - R^{-1}B'\lambda$. Now letting $\lambda = K(t, x)x$, then

$$u = - R^{-1}B'K(t, x)x, \tag{5.64}$$

and

$$\dot{\lambda} = \dot{K}(t, x)x + K(t, x)\dot{x}. \tag{5.65}$$

Replacing $\dot{\lambda}$ and $\dot{x}$ using (5.61) and (5.60) in (5.65) and rearranging and grouping terms,

$$\begin{aligned}&\left[\dot{K}(t, x) + A'(x)K(t, x) + K(t, x)A(x) + Q(x) - K(t, x)BR^{-1}B'K(t, x)\right]x \\ &+ \left(\frac{1}{2}x'\frac{\partial Q(x)}{\partial x}x + \frac{1}{2}x'K(t, x)BR^{-1}(x)\frac{\partial R(x)}{\partial x}R^{-1}B'K(t, x)x\right. \\ &\left.+x'K(t, x)\frac{\partial A(x)}{\partial x}x\right) = 0, K(t_f, x(t_f)) = 0.\end{aligned} \tag{5.66}$$

Equation (5.66) is difficult to solve analytically. However, in the limit, as $t_f \to \infty$, assuming that the (pointwise) state-dependent algebraic Riccati equation (SD-ARE) is satisfied pointwise, it can be assumed that $K(t,x) \rightarrow K(x)$ (i.e., $K(x)$ becomes time independent), and that the SD-ARE will be satisfied, such that in the limit

$$A'(x)K(x) + K(x)A(x) + Q(x) - K(x)BR^{-1}B'K(x) = 0. \tag{5.67}$$

The remaining part of Eq. (5.66) then will also be satisfied in the limit, that is,

$$\dot{K}(x)x + \left(\frac{1}{2}x'\frac{\partial Q(x)}{\partial x}x + \frac{1}{2}x'K(x)BR^{-1}(x)\frac{\partial R(x)}{\partial x}R^{-1}B'K(x)x \right.$$
$$\left. + x'K(x)\frac{\partial A(x)}{\partial x}x\right) \rightarrow 0, \tag{5.68}$$

since the resulting asymptotically stable system with the feedback control $u = -R^{-1}B'K(x)x$ will yield $x \rightarrow 0$ as $t_f \rightarrow \infty$, and thus the necessary conditions for the closed loop solution will approach near optimality.

Stability in Open Left Half Plane

To realize stability by requiring all eigenvalues to be in the open left half plane is insufficient to verify stability, as noted in Sect. 5.2.2 and as shown in the following.

Assuming that the closed loop dynamics of a plant are represented by (5.69), where the control is an SDRE function:

$$\dot{x} = a_c(x), x_1(0) = 2, x_2(0) = 2; \tag{5.69}$$

here $x \in R^2$ is the state vector, $a_c(x) \in R^2$ is a state-dependent vector, and x_1 and x_2 are elements of x, where

$$a_{c,1}(x) = -x_1 + x_1^2 x_2, \tag{5.70}$$

and

$$a_{c,2}(x) = -x_2, \tag{5.71}$$

By factoring

$$a_c(x) = A_c(x)x, \tag{5.72}$$

with

$$A_c(x) = \begin{bmatrix} -1 & x_1^2 \\ 0 & -1 \end{bmatrix}. \tag{5.73}$$

The eigenvalues of (5.73) are (− 1, − 1) for all x. The solution to (5.69) is

$$x_1(t) = \frac{2x_2(t)}{x_2^2(t) - 2}, \tag{5.74}$$

$$x_2(t) = 2e^{-t}. \tag{5.75}$$

Thus as

$$t \to \log\sqrt{2}, \quad x \to \infty, \tag{5.76}$$

which implies that the system is unstable, and with eigenvalues in the open left half plane.

Resistance to Deformation Characteristic

The characteristic as depicted in Fig. 5.52 for material yield stress, as noted in Sect. 5.2.6, is derived from data in [14].

Examples of Eccentricity Compensation

This section presents some simulations of examples[56] for the use of the methods of compensation of eccentricity as described in Sect. 5.2.9. The assumptions used are as follows:

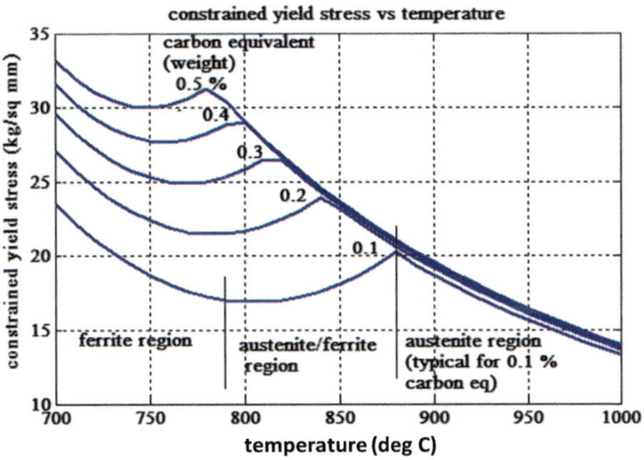

Fig. 5.52 Typical material yield stress characteristic

[56] Examples are based on [8].

- Eccentricity is only in the backup rolls which have duplicate eccentricity; the work rolls are eccentricity free.
- Backup roll diameter is about 1346 mm; mechanical wear is about 2.54 mm or about 0.2%. For simulation and conservatism, a change in diameter of 0.5% is assumed.
- Eccentricity is a sinusoid plus a third harmonic. The period of the fundamental is taken to be one revolution of the backup roll, which corresponds to about 2% of the exit thickness at the mill stand. The peak of the third harmonic is taken to be about 3% of the peak of the fundamental.
- Symbols used are those described in Sect. 5.2.9.

The method used for compensation was the least mean square (LMS) adaptive filtering technique. This technique is well-known and is described in texts on statistical digital processing and adaptive filtering, e.g., [21, 22]. In the simulations a modified algorithm as described in [21] is used to update the filter coefficients. The expression for the algorithm is

$$w_{n+1} = w_n \frac{\beta X_2^*(n)}{\varepsilon + \|X_2(n)\|^2} e_f(n), \tag{5.77}$$

where w is the filter coefficient, n is the discrete time step, β is the gradient step size (normalized), ε is a small positive constant, $X_2(n)$ is the filter input vector, $e_f(n)$ (scalar) is the filter error, $X_2^*(n)$ is the complex conjugate of $X_2(n)$, which (in this case) is $X_2(n)$.

In this simulation an LMS filter of order 12 with a sampling rate of at least 50 samples per period of the sinusoid was assumed for the eccentricity. Several variations were then evaluated to realize some of the characteristics of the controller.

In the initial evaluation the eccentricity $v_1(n)$[57] was taken to be equal to the sinusoid generated by the upper backup roll. The source of the eccentricity was taken based on the measured speed of the upper work roll. Initially no harmonic was assumed. In this instance the filter "learns," and after the learning, the amount of eccentricity was negligible (less than 0.003% of the magnitude of the original eccentricity). Additionally, the learning was completed within two revolutions of the backup roll. Some variations were then each examined as follows:

Variation 1: As previously described in the initial evaluation.
Variation 2: The magnitude of the eccentricity $v_1(n)$ was varied between twice the magnitude of $v_2(n)$ and half the magnitude of $v_2(n)$.
Variation 3: The frequency of the eccentricity $v_1(n)$ was varied between twice the magnitude of $v_2(n)$ and half the magnitude of $v_2(n)$.
Variation 4: $v_2(n)$ is shifted by time steps, i.e., 1, 2, and then 3; each time step is equal to one period of the sinusoid.
Variation 5: A third harmonic was added to the fundamental.

[57] The signals are identified as in Fig. 5.40.

Variation 6: The previous variations were combined to approximately realize the greatest deviation in eccentricity.

The results are that the greatest deviation was less than 0.05% of the peak of the initial eccentricity.

References

1. Cimen T (2010) Systematic and effective design of nonlinear feedback Riccati equation (SDRE) method. Annu Rev Control 34(1):32–51
2. Lewis FL (1986) Optimal control. Wiley, New York
3. Athens M, Falb PL (1966) Optimal control, an introduction to the theory and its application. McGraw Hill, New York
4. Sarsembayev B et al (2020) SDRE-based integral sliding mode control for wind energy conversion systems. IEEE Access 8:51100–51113
5. Roudkenary KA et al (2016) SDRE control of non-linear affine systems. In: IEEE 4th international conference on control and automation, Qazvin, Iran, pp 239–244
6. Vidyasagar M (1978) Nonlinear system analysis. Prentice-Hall, Englewood Cliffs
7. Erdem EB (2001) Analysis and real-time implementation of state-dependent Riccati equation controlled systems. Ph.D. thesis, University of Illinois at Urbana–Champaign, Champaign
8. Pittner J, Simaan MA (2011) Tandem cold metal rolling mill control, chap 5. Springer, London
9. Pittner J, Simaan MA (2010) Quality improvement of tandem hot metal strip rolling using an augmented state-dependent Riccati equation technique. In: 49th IEEE conference on decision and control, Atlanta, Georgia, pp 5180–5185
10. Kinney CS, Laub AJ (1995) The matrix sign function. IEEE Trans Automat Control 40(8):1330–1348
11. Simaan MA, Pittner J (2015) Mill control system and method for control of metal strip rolling. US patent 9,905,886
12. Pittner J, Simaan MA (2014) Control of the tandem hot strip mill under expanded uncertainties and disturbances. IEEE Trans Ind Appl 50(5):3085–3094
13. Pittner J, Simaan MA (2015) Advanced control to reduce the likelihood of cobbles in the tandem rolling of hot metal strip. IEEE Trans Ind Appl 51(5):4305–4312
14. Herman JC, Leroy V (1997) New possibilities of the ferrite hot rolling applied to low carbon steels. In: TechCo 29 conference, Seoul, Korea
15. Pittner J, Simaan MA (2016) Use of advanced control with virtual rolling to improve the control of the threading of the tandem hot metal strip mill. In: Proceedings of 2016 IEEE industry applications society annual meeting, Portland, OR, pp 1–8
16. Steinboeck A et al (2014) Control of strip tension in a rolling mill based on loopers and impedance control. In: Proceedings international federation automatic control, Cape Town, South Africa, pp 10646–10651
17. An BJ et al (2001) Tension control system for hot strip, mills. In: IEEE international symposium industrial electronics, Pusan, Korea, pp 1452–1457
18. Imanari H et al (1997) Looper H-infinity control for hot strip mills. IEEE Trans Ind Appl 43(3):790–796
19. Pittner J, Simaan MA (2017) Use of fault tolerant techniques with virtual rolling to improve the robustness to measurement faults in the control for tandem hot metal strip rolling. In: Proceedings of 2017 IEEE industry applications society annual meeting, Cincinnati, OH, pp 1–8
20. Yang X et al (2016) Online monitoring system design for roll eccentricity in rolling mills. IEEE Trans Ind Electron 63(4):2559–2568
21. Hayes MH (1996) Statistical digital signal processing and modeling, chap 9. Wiley, New York

22. Widrow B, Stearns S (1986) Adaptive signal processing. Prentice-Hall, Englewood Cliffs
23. Pittner J, Simaan MA (2013) An advanced control method for saving energy in the tandem rolling of hot metal strip. In: 2013 IEEE Energytech, Cleveland, OH, pp 1–6
24. Industry statics (2022). Iron Steel Technol 19(10):11–18
25. Roberts WL (1983) Hot rolling of steel. Marcel-Dekker, New York
26. Malloci I et al (2010) Robust steering control at hot strip mill. IEEE Trans Control Syst Technol 18(4):908 917
27. Zhou K, Doyle DC (1988) Essentials of robust control. Prentice-Hall, Upper Saddle River
28. Prinz K et al (2017) Automatic gauge control under laterally asymmetric rolling conditions combined with feedforward. IEEE Trans Ind Appl 53(3):2560–2568
29. Prinz K et al (2018) Optimization-based feedforward control of the strip thickness profile in hot strip rolling. J Process Control 64:100–111
30. Rawlings JB (2000) Tutorial overview of model predictive control. IEEE Control Syst Mag 20(3):38–52
31. Allgöwer F et al (1999) Nonlinear predictive control and moving horizon estimation-an introductory overview. In: Frank PM (ed) Advances in control highlights of ECC99. Springer, London
32. Yin F (2020) Discrete model predictive control scheme for an integrated gauge-looper control system in a tandem hot strip mill. IEEE Access 8:73972–73985
33. Furlan R et al (2008) Friction compensation in the interstand looper of hot strip mill: a sliding-mode control approach. Control Eng Pract 16(2):214–224
34. Zhong Z et al (2012) Looper-tension sliding mode control for hot strip finishing mills. J Iron Steel Res 19(1):23–30
35. Feng Y et al (2014) Chattering free full-order sliding mode control. Automatica 50:1310–1314
36. Du H et al (2016) Chattering-free discrete-time sliding mode control. Automatica 68:87–91
37. Young DK (1999) A control engineer's guide to sliding mode control. IEEE Trans Control Syst Technol 7(3):328–342
38. Spurgeon SK (2014) Sliding mode control: a tutorial. In: ECC2014. Kent Academic Repository, University of Kent, UK, Strasbourg, France
39. Green M, Limebeer DIN (1995) Linear robust control. Prentice-Hall, Upper Saddle River
40. Pittner J, Simaan MA (2011) Tandem cold metal rolling mill control, chap 4. Springer, London
41. Geddes EJM, Postlewaite I (1998) Improvements in product quality in tandem cold rolling using robust multivariable control. IEEE Trans Control Syst Technol 6(2):257–269

Chapter 6
Main Drives and Motors

6.1 Introduction

An objective of this chapter is to present a very basic introduction to the characteristics of variable speed drives and motors as they pertain to the control of tandem hot strip rolling. It is hoped that the material presented[1] will provide an easier basis for understanding the interfacing between control and drives so that the ultimate objectives in performance can be readily realized. What follows will be based on typical tandem hot strip mill applications; specifically, voltage source converters and cycloconverters in the medium voltage range as drives for synchronous motors and for cage rotor induction motors.[2] Additionally, the concept of closed loop field-oriented control will be addressed. Some basic fundamentals will be presented with the further objective of helping the reader to better understand certain issues as described in various texts, archival literature, and manufacturer's publications.

6.2 The Cage Rotor Induction Motor, Multi-phase

The multi-phase cage rotor induction motor is used in an almost endless variety of industrial applications as an actuator that is simple, rugged, economical, and whose speed is nearly constant when supplied by a constant frequency source. The motor consists of a rotor and a stator. The rotor is supported on bearings, while the stator is

[1] Copyright 2011 Springer Nature. Figures in this chapter are adapted and reprinted, by permission, from [1].

[2] Most of the main drive motors are synchronous machines as they are more capable of handling the higher loading requirements. However, there are some interests in using cage rotor induction machines for certain applications, and some have been successfully applied. Additionally, an initial basic understanding of the characteristics of the widely familiar cage rotor machine helps to better understand the basic characteristics of the synchronous machine.

J. Pittner and M. A. Simaan, *Tandem Hot Metal Rolling Mill Control*, Advances in Industrial Control, https://doi.org/10.1007/978-3-031-58082-6_6

stationary. Alternating currents exist in the stator windings and in conductive bars of aluminum or copper in slots in the rotor. In the case of a wound rotor motor, windings are inserted in the slots and brought out via mechanical brushes and slip rings.

The stator windings are connected to a multi-phase source, while the currents in the rotor bars are induced, as inferred by the motor's name. Most of the motors are three phase, but some are of more phases that are used to satisfy certain special requirements. It also can be noted that the motor can operate as a generator, wherein power is supplied to other entities, although this generally is not the case for tandem hot strip mills. Some additional characteristics of the motor are as follows:

- The stator and rotor are constructed using thin steel sheets of a special grade for motor usage, and treated to reduce the undesirable effects of eddy currents.
- In a cage type rotor, the rotor bars are short circuited at each end, i.e., there is no connection of the rotor bars to other circuits. In the wound rotor type of motor, the windings are brought out via mechanical brushes and slip rings.
- In the stator the windings are arranged to produce an approximately sinusoidal spatial distribution of a magnetomotive force, denoted as the mmf. In a three-phase machine, the coils are arranged as denoted functionally as in Fig. 6.1, where each of the three coils is excited by a phase of a balanced three-phase source.[3] In Fig. 6.1 the three coils are denoted as a–a', b–b', and c–c'. The coils are shown as concentrated for simplicity, but in actuality are distributed to produce a nearly sinusoidal spatial mmf.

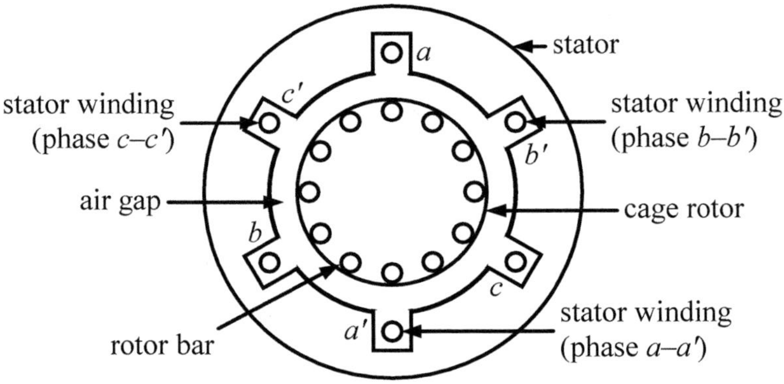

Fig. 6.1 Section of multi-phase motor, depicting cage rotor and stator in a two pole machine[4]

[3] More detail on the fundamentals of induction motors is available in texts, e.g., [2, 3].

[4] Copyright Springer Nature 2011, Fig. 6.1. Portions are cosmetically reprinted, with permissions, from [1].

6.2.1 Magnetic Field Characteristics

A magnetic field that rotates is produced when the stator windings are connected to a balanced three-phase source. It is important to understand how this field is produced to better understand the operation of the multi-phase machine.[5]

In the case of a two pole machine (Fig. 6.1)[6] the windings around the stator are displaced from each other by 120°. The electrical angular separation is not always the same as the spatial separation (i.e., the mechanical angle). The relation between the mechanical angle and the electrical angle is based on one cycle of flux distribution as described in (6.1):

$$\theta_{\mathrm{m}} = \frac{2}{P}\theta_{\mathrm{e}}, \tag{6.1}$$

where θ_{e} is the electrical angle that is based on the one cycle of the flux distribution around the air gap, where one cycle is 360 electrical degrees, and P is based on the number of wavelengths per revolution. In a two pole machine the electrical and mechanical angles are equal to each other.

For initial insight, an mmf wave (F_{a}) is taken to be produced by a time invariant current in coil *a–a′* in Fig. 6.1. The peak $F_{\mathrm{a(pk)}}$ of the coil is produced by the peak in the current in coil *a*. Coils *b* and *c* are located on the stator as in Fig. 6.1; these are located around the air gap at angles 120 and 240° as measured counter-clockwise from coil *a*. Thus, the total mmf wave can be described as:

$$F(\theta) = F_{\mathrm{a(pk)}}\cos(\theta) + F_{\mathrm{b(pk)}}\cos(\theta - 120^\circ) + F_{\mathrm{c(pk)}}\cos(\theta - 240^\circ). \tag{6.2}$$

Now when the windings are excited by a balanced three-phase source, the instantaneous peak of the mmf in a winding is proportional to the instantaneous peak in the current in the winding, so that the mmf in each winding is as follows:

$$F_{\mathrm{a(pk)}} = F_{\mathrm{a(max)}}\cos(\omega t), \tag{6.3}$$

$$F_{\mathrm{b(pk)}} = F_{\mathrm{b(max)}}\cos(\omega t - 120^\circ), \tag{6.4}$$

$$F_{\mathrm{c(pk)}} = F_{\mathrm{c(max)}}\cos(\omega t - 240^\circ), \tag{6.5}$$

where ω is the frequency of the sinusoidal current in a winding. In can be shown (Appendix Section "Derivation of Angular Velocity of MMF Vector") that the result

---

[5] The operation of single-phase machines is different. These are used for much lower power applications. More detail on single-phase machines can be found in applicable texts.

[6] A two pole machine is used as a simpler example to provide some insight into the basic fundamentals. These can be extended for machines with greater numbers of poles, as is common in certain drives.

is an mmf vector rotating around the air gap at an angular velocity ω (i.e., in electrical radians per second).

What is shown previously is based on a two pole machine. This can be extended to higher order poles by locating a two pole configuration serially around the stator of the machine. Thus for any machine, the rotational velocity ω_m of the mmf wave is as in (6.6) with respect to the stator, ω_e is the frequency of the electrical angle of the flux distribution, both in rad/s, where

$$\omega_m = \frac{2}{P}\omega_e, \tag{6.6}$$

and

$$\eta = \frac{120f}{P}, \tag{6.7}$$

and where P is the number of poles in the machine, f is the frequency in Hertz of the three-phase source, η is the actual rotor speed, η_1 is the synchronous speed, both in in rev/min, and s is the per unit slip:

$$s = \frac{\eta_1 - \eta}{\eta_1}. \tag{6.8}$$

The frequency of currents in the rotor is:

$$f_r = sf, \tag{6.9}$$

where f is the frequency of the stator currents. It can be noted that at a start from a stationary rotor, the slip is 1.0, and the frequency of the rotor currents is the same as the frequency of the stator currents.

6.2.2 Torque

To understand how a cage rotor machine produces torque, it is helpful to understand how the air gap flux density waves and rotor mmf waves interact. For additional insight, a description of this is described in Appendix Section "Interaction of Flux Density Waves and Rotor MMF Waves to Produce Torque". Based on what is described therein, a speed–torque relationship for a typical cage rotor machine is depicted in Fig. 6.2.

In this figure it can be noted that the machine is being operated as a motor at its nominal frequency and rated terminal voltage. It should be noted that, the frequency of the stator is fixed; however, during deceleration the frequency of the stator and its terminal voltage are continually changing.

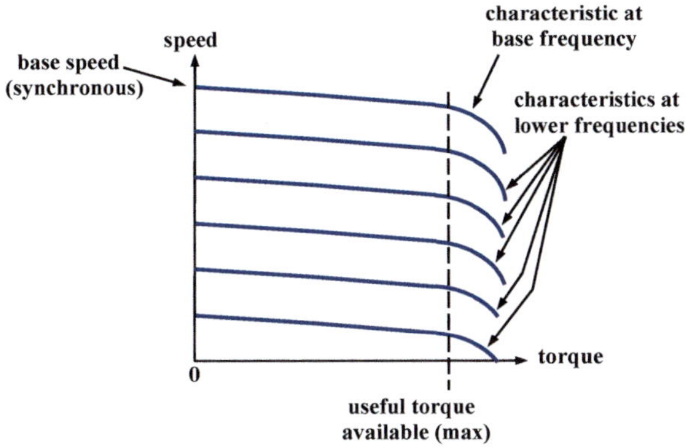

Fig. 6.2 Typical cage rotor induction motor characteristics of torque and speed, at and below base frequency[7]

The rotor torque T_r can be expressed[8] as:

$$T_r = K_r \left(\frac{P}{2} \right)^2 B_{pk} F_{r,pk} \sin(\beta_r), \tag{6.10}$$

where K_r is a constant depending on rotor geometry, P is the number of poles, B_{pk} is the peak of the flux density wave, $F_{r,pk}$ is the peak of the mmf wave generated by the rotor currents, and β_r is the angle (electrical) between the flux density wave and the mmf wave. A useful approximation is:

$$T_r = K I_r \sin(\beta_r), \tag{6.11}$$

where K is a constant, and I_r is proportional to the root-mean-square of the rotor current.

6.2.3 *Regeneration*

In general, induction motor drive regeneration is usually less necessary in tandem hot mills, except for certain special cases. However, to present a somewhat more complete basic description of the cage motor characteristic, a brief description of regeneration is included as follows.

The motor regenerating implies that the motor is supplying power to other elements as loads, such as motors, or the plant power grid. If the induction motor is to regenerate, the speed of the load coupled to the rotor must be greater than the

---

[7] Copyright Springer Nature 2011, Fig. 6.2. Portions are cosmetically reprinted, with permissions, from [1].

[8] Derivation is in [2].

synchronous speed of the rotor. If this is so, the rotor current, the rotor pole polarities, and the torque direction will be reversed from their values when motoring.[9]

6.2.4 Variable Speed

To help understand the motor characteristics during the changes in the frequency of the power source to realize a change in speed, two areas of interest are the operation at and below base frequency, and the operation above base frequency. As depicted in Fig. 6.2, the operation at and below base frequency is essentially a constant torque area; the operation above base frequency is essentially a constant horsepower area.[10] In what follows, open loop control is assumed to realize the machine characteristics under each of these conditions. For operation at and below base frequency an important entity in the cage rotor machine is the air gap flux, which is dependent on the air gap voltage and the power source frequency:

$$\varnothing = k_{\mathrm{v}} \frac{v_{\mathrm{ag}}}{f}, \tag{6.12}$$

where

$$V_{\mathrm{t}} = V_{\mathrm{r}} + V_{\mathrm{x}} + V_{\mathrm{ag}}, \tag{6.13}$$

and where $\varnothing$ is the airgap flux, V_{ag} is the air gap voltage, f is the frequency of the power source, k_{v} is a constant, V_{t} is the terminal voltage of the machine, and V_{r} and V_{x} are the voltages across the stator resistance and leakage reactance. At the base frequency and with the rated terminal voltage applied, the machine will run as described in its base speed characteristic (Fig. 6.2).

Now if the machine speed is to be reduced, to avoid magnetic saturation the air gap flux should be kept very nearly constant. This done by a proportional reduction in the frequency of the power source and the air gap voltage.

A proportional reduction is closely realized since the magnitudes of V_{r} and V_{x} are quite small with respect to V_{ag} over most of the speed range. However, at near the lower part of the speed range, the magnitudes of the voltages V_{r} and V_{x} are no longer small with respect to the air gap voltage V_{ag}, so that the constant torque is reduced. Thus a compensating terminal voltage is applied at the lower end of the speed–torque characteristic to keep a nearly constant air gap flux at the lower speeds.

Therefore, for the motoring operation, the constant torque characteristic is reasonably retained at and below the base frequency. In the case of the regenerating operation below the base frequency a compensating voltage is applied and a similar characteristic is realized.

---

[9] More detail is in [1–3].

[10] Additional detail is in [1–3].

For operation above base frequency, the voltage at the motor terminals is held constant while the frequency is increased, which results in constant horsepower with decreasing torque [3]. The airgap flux decreases in accordance with (6.12), with the resulting motoring characteristics. In the case of regeneration, the characteristic would be similar except that, as noted previously, the rotor current, the rotor pole polarities, and the torque direction will be reversed from their values when motoring.

6.3 The Synchronous Motor

In general, the synchronous motor is usually preferred over the cage rotor induction motor as the main drive for hot strip mill stands. This is because the synchronous motor is more suitable for applications wherein there are heavy torque demands, higher efficiency, a constant speed, and unity power factor that can be realized by adjusting the field excitation.

The construction of the stator of the motor is similar to that of the stator of the induction motor (Fig. 6.1); i.e., the stator windings are placed in slots arranged around the surface of the stator such that a sinusoidal spatial distribution of the magnetomotive force (mmf) in the air gap is realized. The rotating field of the rotor induces a voltage in the stator windings, which are identified as the armature of the motor. The stator windings are connected to an external multi-phase source. However, unlike the induction motor, the rotor windings are excited by a separate direct current source, rather than by the rotation of the air gap flux. The current for the rotor is obtained via an excitation system which is mounted on the rotor shaft, without needing slip rings and brushes.

The rotor can have salient poles or non-salient poles. In the case of non-salient poles, the rotor (denoted as a cylindrical rotor) has slots in its surface which contain conductors, similar to the rotor of the induction motor. The cylindrical rotor results in a uniform air gap with a sinusoidal mmf distribution with same number of poles as in the stator winding. In general, the cylindrical rotor has been preferred as it has higher mechanical strength and reliability, less windage losses, plus other advantages. As in the case of the induction motor, the synchronous motor can also operate as a generator to regenerate power to other loads in the mill or the plant.[11]

6.3.1 *Torque*

An understanding of the interaction of the production of torque by the interaction of the flux and mmf waves is important to understand the synchronous motor. To help in this effort, it is assumed that the stator is excited by a balanced three-phase source of constant frequency. This results in a magnetic field in the air gap rotating

[11] More detail on the synchronous motor is available in [2, 3] and manufacturer's literature.

at synchronous speed, similar to the rotating magnetic field described in Sect. 6.2.1 for the stator of the induction machine.

Additionally, the following are assumed:

- the losses and saturation effects can be neglected, such that the stator reactance and resistance are zero;
- the rotor is turning at synchronous speed;
- the position of the rotor and the current in its conductors are such that the voltages that are induced in the stator (i.e., the armature) are equal to the voltages in the three phases of the stator source;
- thus the currents in the stator are zero, and the rotor torque (as described in (6.14)) is zero:

$$T_{\mathrm{m}} = K_{\mathrm{t}}\left(\frac{P}{2}\right)^2 \phi_{\mathrm{rt}} F_{\mathrm{rt}} \sin(\delta_{\mathrm{rf}}), \tag{6.14}$$

where:

- T_{m} is the rotor torque, similar to (6.10);
- K_{t} is a constant;
- P is the number of poles;
- ϕ_{rt} is the air gap flux wave;
- F_{rt} is the rotor current mmf wave; and
- δ_{rf} is the angle between the mmf wave and the flux wave.

If now some load torque (T_{l}) is applied to the shaft, the rotor will move backward with respect to the air gap flux wave, and the torque T_{m} will move in a direction (ω) opposite to the applied load torque. After some transients the torques will balance, such that $T_{\mathrm{m}} + T_{\mathrm{l}} = 0$; the rotor is at synchronous speed, and there are currents I_{α} in the armature windings. These currents will produce an mmf field, denoted as the armature reaction, which adds to the mmf field of the rotor to result in an mmf field. This then adjusts the air gap flux to keep the voltages induced in the armature equal to the source voltages.

It can be noted that if the load torque is continually increasing, the angle δ_{rf} will keep increasing until an angle of 90°. This angle is the maximum torque that can be realized at synchronous speed. If the maximum torque is exceeded the rotor will reduce speed and drop out of synchronism.

In the case of regeneration, the load torque is reversed so that the machine becomes a generator, similar to the characteristic of the cage rotor machine.

6.3.2 *Variable Speed*

The two areas of operation to be addressed are (1) at and below base speed, and (2) above base speed, wherein base speed is the speed of the machine with its rated voltage and frequency at its terminals. Operation at and below base speed is similar to the cage rotor machine wherein the objective is to maintain a nearly constant torque characteristic over the speed range. Variable speed operation above base speed can be unnecessary for certain mills, but it can be used wherein there is a requirement for this function in particular applications such as zooming[12] of the mill. In these applications, the speed is increased above base speed with an essentially constant horsepower characteristic. What follows examines the machine characteristics in each of these two areas. In each case open loop control is assumed.

In each case the torque of the machine is denoted as in (6.14), and also can be described as in [4], i.e.,

$$T_{\mathrm{m}} = K_{\mathrm{tm}} \frac{v_{\mathrm{tm}}}{f} \sin(\delta_{\mathrm{rf}}), \tag{6.15}$$

where T_{m} is the machine torque, K_{tm} is a constant, V_{tm} is the machine terminal voltage, f is the frequency of the power source, and δ_{rf} is the angle between the mmf wave and the flux wave in the machine as in (6.15). However, in [4] the resistance portion of the stator impedance is taken to be zero at the base frequency, as it is quite small when compared to the reactive component. The maintenance of constant torque is realized by adjusting the terminal voltage of the machine to be proportional to the frequency, similar to what is done in the cage rotor induction motor. An additional adjustment in the terminal voltage is applied at the lower end of the frequency range to compensate for the resistive portion of the stator impedance, since the reactive portion is small at the lower frequencies. Figure 6.3 depicts the characteristic for motoring and generating applications.

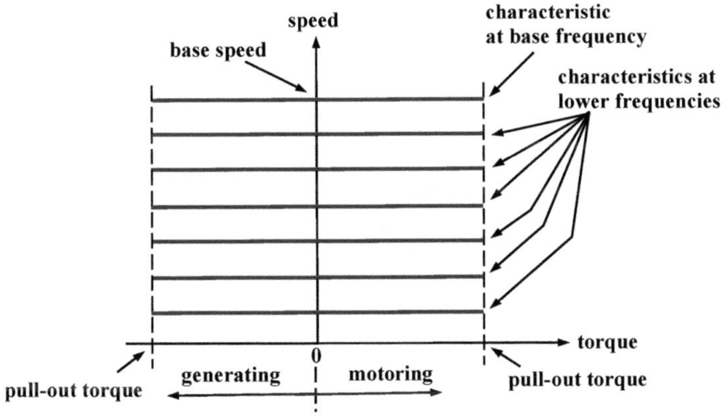

Fig. 6.3 Synchronous machine characteristics at and below base frequency[13]

[12] Zooming is a significant increasing of the mill speed in response to certain events possibly occurring during operation.

[13] Copyright Springer Nature 2011, Fig. 6.3. Portions 3 are cosmetically reprinted, with permissions, from [1].

At above base speed the frequency is increased with the terminal voltage remaining constant. In this case the torque decreases as the speed increases and the horsepower remains constant.

6.4 Drives for Main Motors

In this section some very basic concepts of power electronic circuitry and associated control methods of the main drives of the tandem hot strip mill are addressed. The objective is to provide a background to help understand some of the main concepts that are the basis for some successful applications and modern advancements in this area. What follows will consider the drives that are for synchronous motors and cage rotor induction motors.[14]

6.4.1 Voltage Source Converter

Typical of present equipment that controls power to the drive motor is depicted functionally in Fig. 6.4. The three main parts of the drive are the active front end, the DC link, and the motor inverter. This has been used successfully in many applications and is generally a preferred method.

6.4.1.1 Active Front End and DC Link

The active front end provides power to the bridge arrangement of large capacitors which are a link to the semiconductor inverter, and also provides the full motoring and generative capability of the drive. In most recent applications the semiconductors are predominately insulated gate bipolar transistors (IGBTs), or in certain products, integrated gate commutated thyristors (IGCTs).[15] The gating actuation of the IGBTs is such that the harmonics introduced into the power system are quite low, which is realized by controlling the switching of the IGBTs to maintain a nearly sinusoidal line current.

Also, brief fluctuations in the line voltage are reasonably tolerated since the DC link voltage is held nearly constant. This is because the capacitors are quite large so as to provide a stiff voltage source to support the operation of the inverter.[16] Thus the

[14] What is presented herein is intended to be examples for the development of basic insight. More specific detail for further insight using possible alternate viable methods is available via associated archival technical literature (e.g., [5]) and applicable recent manufacturers' publications.

[15] Detailed usage of IGCTs versus IGBTs by certain drive manufacturers is usually available in their publications. What is described herein for IGBTs is generally applicable to both.

[16] The capacitor circuit is split to produce a neutral point that is needed by the inverter for its proper operation.

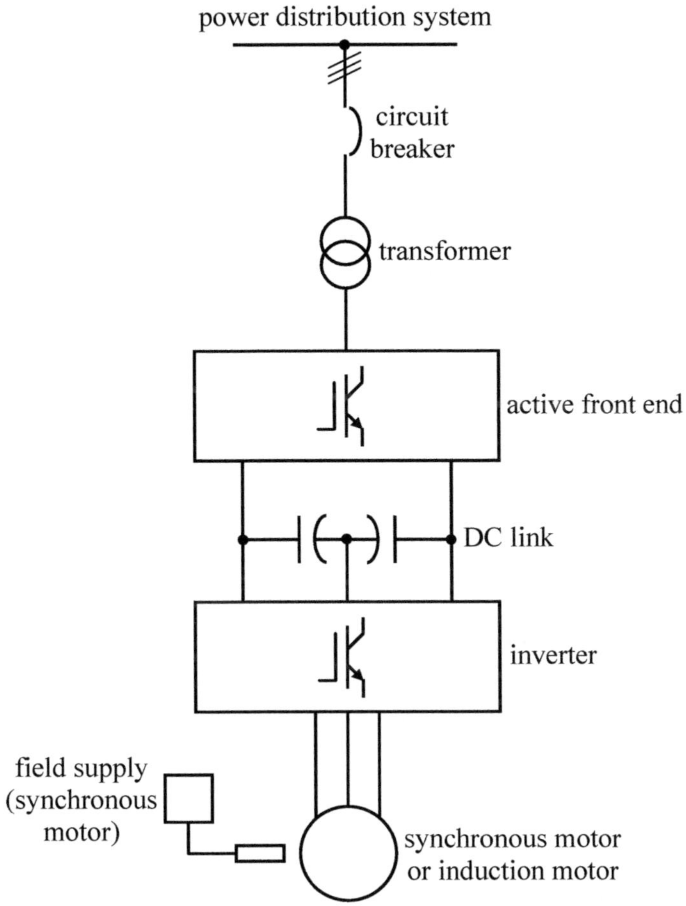

Fig. 6.4 Power functional schematic for voltage source converter[17]

drive system is less sensitive to these voltage fluctuations, especially to brief dips in the line voltage. Further, a power factor of 1.0 can be realized since the reactive power can be controlled; or the power factor can be adjusted to compensate for variations in other loads connected to the same bus as the drive.

6.4.1.2 Inverter

The inverter is an arrangement of power semiconductors that control power to the motor. The inverter converts the voltage of the DC link into a waveform for variable

[17] Copyright Springer Nature 2011, Fig. 6.4. Portions are cosmetically reprinted, with permissions, from [1].

speed operation of the drive motor for motoring or regenerating over its full speed range. In this case as an example for basic insight, the inverter consists of IGBTs in a three level three-phase arrangement with a neutral point. The neutral point is provided in the DC link by the arrangement of the capacitors.

The inverter is pulse width modulated (PWM) wherein broken-up square waves are applied based on a fundamental output voltage. Each of the square waves is broken into pulses wherein the fundamental is a sinusoid. The width of the pulses is adjusted to determine the magnitude of the fundamental, while the frequency of the square waves is determined by the desired frequency of the fundamental.

An advantage of the PWM technique is that larger and costly filters are not needed to reduce lower order harmonics in the waveform of the current. Additionally, in the case of the PWM, the output voltage is well-controlled. Moreover, the commutation losses of the inverter are quite low which helps the overall efficiency of the drive to be quite high (about 99%), and the output current has very low harmonic current which reduces pulsations to the motor shaft and removes any need for the derating of the drive motor. Thus consideration of the characteristics of the PWM technique and the actual application, would make it a desirable and preferred drive.

6.4.2 Cycloconverter

In general, the cycloconverter is used for very large loads in industrial applications. However, in many cases the cycloconverter also might be considered for use in the tandem hot strip mill, such as in gearless coupling between the motor shaft and the work roll. Thus for completeness, what follows is a brief description of the cycloconverter.

The cycloconverter is a circuit of thyristors that converts the frequency of an input power source to another frequency at an output load. When used as a power source for a hot strip finishing mill main drive motor, the cycloconverter adjusts the voltage and frequency of the stator windings of the motor. This is done by the control of dual converters, each of which is a set of thyristors that is applied for each individual phase of the motor.

In a dual converter the thyristor circuits are arranged to provide a DC output from an alternating current source such as the plant power system. The DC output then is varied to provide an alternating current that is close to sinusoidal and that is at a lower frequency than the input source. However, there is an upper limit on the frequency that can be realized with a reasonably sinusoidal waveform; in general, this is about 20% of the input frequency.

In the cycloconverter it is possible to realize a control that results in a low harmonic content in the motor torque. Nonetheless it can be noted that, even though the harmonic content is lower than with a DC drive, the harmonics are more difficult to filter as they have a non-integer characteristic. This is due to the continuous changing in the firing angle of the thyristors.

Additionally, the frequency of the output is much lower than that for frequencies of other drives such as the voltage inverter. This requires a much larger motor which can result in a cost increase over other drives. Since the frequency and voltage are lower, the best power factor is about 0.8. The losses of the cycloconverter are fairly low, as the efficiency is about 99%. However, there is a higher harmonic content on the incoming line. Thus in general and considering the requirements of a particular application, the cycloconverter should be carefully considered before selecting it as a desirable choice for a drive.[18]

6.4.3 *Control of Drives*

Drive controllers in use for cage rotor induction and synchronous machines are often based on field-oriented (i.e., vector) control.[19] This method is very desirable in that it is similar to one that has been highly successful in use for many earlier applications in control of a separately excited DC motor.[20] A very brief review of the control of the DC machine will help to better understand this technique when applied to cage rotor induction and synchronous motors.

In a separately excited DC machine, the shaft torque is described as in (6.16):

$$T_{\mathrm{m}} = K\Phi_{\mathrm{f}} I_{\mathrm{a}}, \tag{6.16}$$

where T_{m} is the motor torque, ϕ_{f} is the flux resulting from the current in the machine shunt field, I_{a} is the armature current, K is a constant, and wherein saturation and armature reaction can be neglected. In the DC machine the field resulting from the armature current is orthogonal to the shunt field and both are stationary in space. Because of the orthogonality, the action of the armature current has negligible effect on the shunt field flux, so that a desired rapid response to a change in the torque can be realized by a similar rapid change in the armature current, without affecting the shunt field current. And also if the shunt field current is changed, the armature current remains unaffected, and the machine torque is affected. Thus with suitable control techniques, similar characteristics can be realized in the cases of the cage rotor induction and synchronous motors.

6.4.3.1 Control of the Cage Rotor Induction Machine

In the case of earlier scalar control, an attempt to change the torque by adjusting the frequency to change the flux, and with only a corresponding adjustment in the terminal voltage, has resulted in a less desirable response in the torque. To improve in

---

[18] More detail on dual converters and cycloconverters is available in [5, 6].

[19] Often vector control is denoted as orthogonal or decoupling control.

[20] A typical characteristic, similar to a DC motor, is described in [1].

this, a method denoted as field-oriented control, also better known as vector control was developed.

In this method, machine variables are transformed into a reference frame that rotates at synchronous speed. In this frame the flux and current vectors are expressed as orthogonal entities. By the use of coordinate changes in this frame, the currents in the machine stator windings can be made to be expressed in a rotating expression for torque that is similar to what is seen in a DC machine, as in (6.17):

$$T_r = K\Phi_{r,pk} I_{s,q}, \tag{6.17}$$

where T_r is the rotor torque, $\phi_{r,pk}$ is the peak value of the rotating machine flux vector ϕ_r that is produced by the flux producing component $I_{s,d}$ of the stator current I_s, $I_{s,d}$ is in phase with ϕ_r; $I_{s,q}$ is the component of the stator current that is orthogonal to ϕ_r and $I_{s,d}$; and K is a constant. Thus the increase in $I_{s,q}$ will produce an increase in the torque T_r without changing the flux ϕ_r, i.e., an increase in the torque producing component in the stator current increases the torque without increasing the flux; and that an increase in the flux producing component in the stator current increases the torque and without producing an increase in the torque producing portion of the stator current.[21]

In Fig. 6.5 a functional description of a field-oriented controller is presented to help provide a basic understanding of the control.[22] The transient response of the control is very much like that of a DC machine. This vector control method has been quite useful for drives in control of the tandem hot strip mill.

6.4.3.2 Control of the Synchronous Machine

Similarly, as in the case of the cage rotor induction machine, the transient response of the synchronous machine can be significantly improved by the use of field-oriented (vector) control. In the case of the synchronous machine the transient response is adversely affected by the slow rise in the rotor field current. The method to improve the response is to add magnetizing current to increase the flux during transient situations. This is done as depicted in Fig. 6.6a, b by use of a concept quite similar to the drive for the cage rotor induction machine. However, there is a difference, which is the control of the rotor field current.

In the control of the synchronous machine, variables are described in a reference frame that rotates at synchronous speed. In this frame the flux and current variables are arranged orthogonally in the rotating frame, similar to the control of the cage rotor induction machine. In the synchronous machine the flux at steady state is set by the current in the rotor windings. However, during a transient condition, the

[21] It is assumed for simplification that the voltage across the rotor reactance is negligible.

[22] Additional detail of a specific controller is outside the intended scope of this section. Some examples, such as the detail of the reference frame, are available in applicable academic literature, e.g., [5]. Likewise with the control of the synchronous machine.

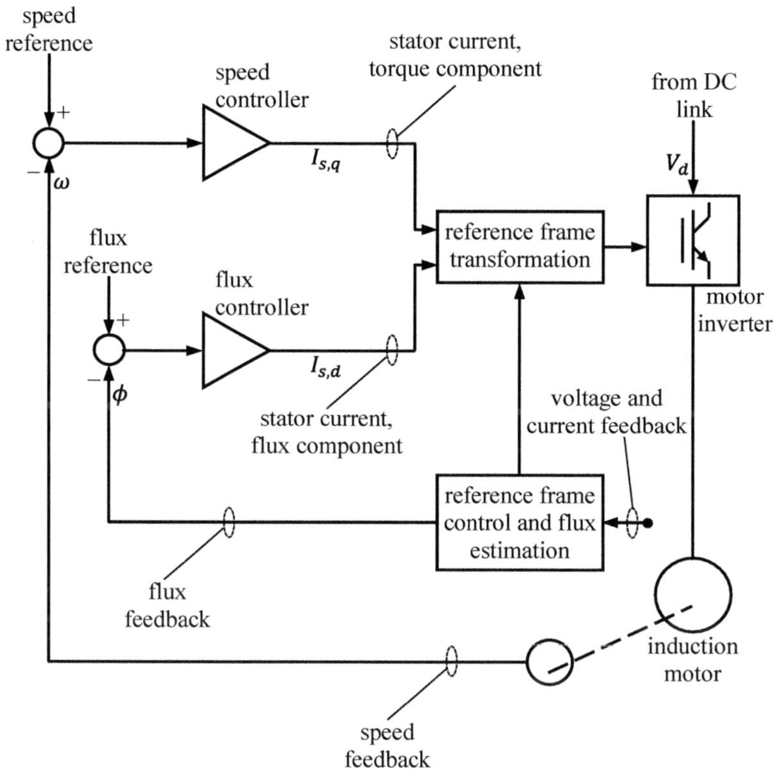

Fig. 6.5 Functional description of a field-oriented control of a cage rotor induction motor[23]

slower change in the current in the rotor windings results in a slower response in the controller.

At a steady-state condition of the rotor, the relationship between the magnetizing current $I_{\mathrm{m,r}}$ and the rotor field current $I_{\mathrm{f,r}}$ is as noted in (6.18):

$$I_{\mathrm{m,r}} = I_{\mathrm{f,r}} \cos\left(\delta_{\mathrm{mf,r}}\right), \tag{6.18}$$

where the reference $I^*_{\mathrm{f,r}}$ to the field current controller is related to $I^*_{\mathrm{m,r}}$ as in (6.19):

$$I^*_{\mathrm{f,r}} = \frac{I^*_{\mathrm{m,r}}}{\cos\left(\delta_{\mathrm{mf,r}}\right)}, \tag{6.19}$$

and where in (6.18) $\delta_{\mathrm{mf,r}}$ is the relation between the magnetizing current and the field flux.

---

[23] Copyright Springer Nature 2011, Fig. 6.5. Portions are cosmetically reprinted, with permissions, from [1].

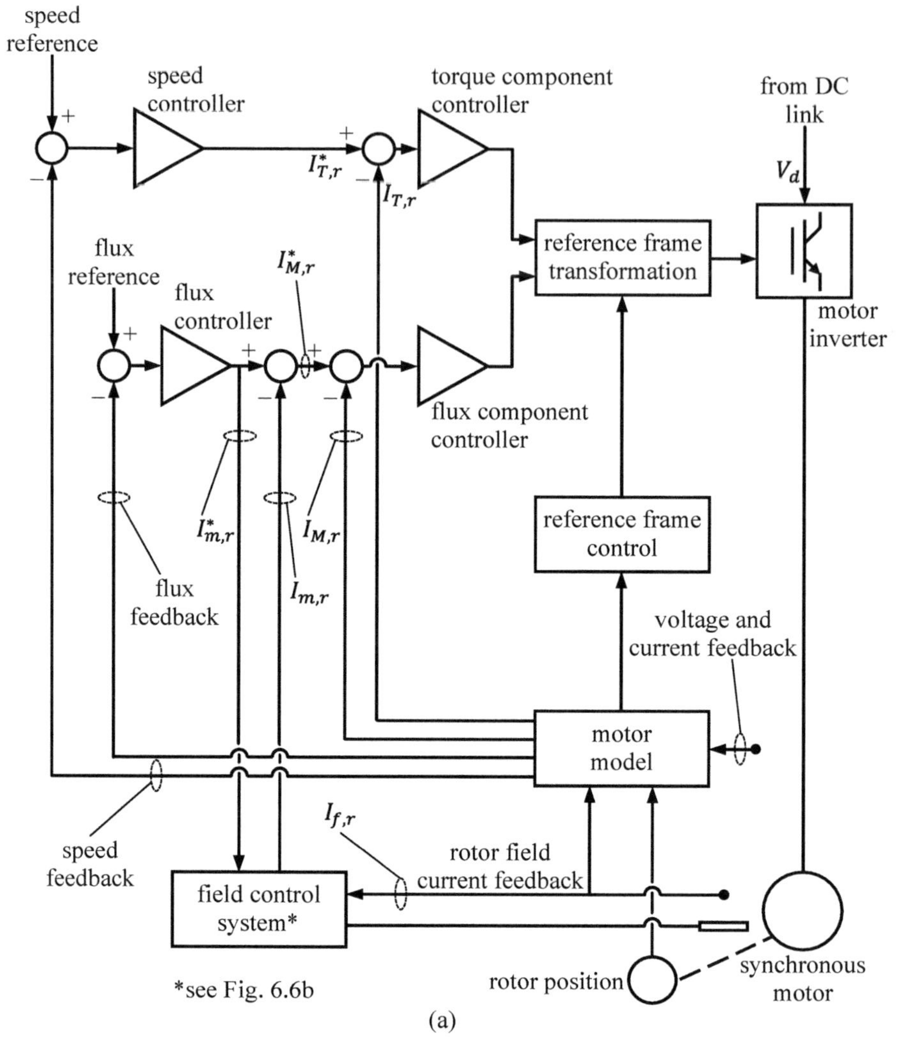

Fig. 6.6 **a** Functional description of a field-oriented control of a synchronous motor. **b** Functional description of the field control system of a synchronous motor (supplements (**a**))[24]

At a steady-state condition of the synchronous machine and operating at unity power factor, the angle between the stator voltage and current is zero. In this case (Fig. 6.6a) the two orthogonal components of stator current are the torque component $I_{T,r}$, and the flux component $I_{M,r}$ which is zero. In this situation, the flux is established by the rotor field current $I_{f,r}$ such that the reference $I^*_{m,r}$ is balanced by $I_{m,r}$ which is

[24] Copyright Springer Nature 2011, Fig. 6.6a and b Portions are cosmetically reprinted, with permissions, from [1].

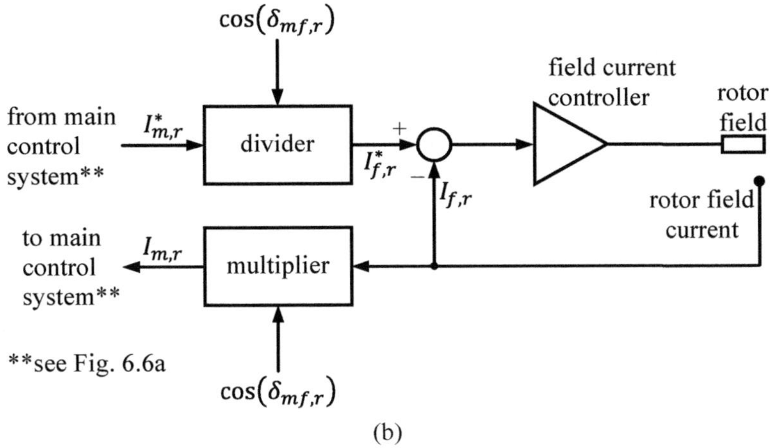

Fig. 6.6 (continued)

determined totally by the field current $I_{f,r}$ such that $I^*_{M,r}$ is zero and thus $I_{M,r}$ is zero by control action.

In the case of a transient condition with the machine operating in the constant torque region (e.g., Fig. 6.3), it is assumed that a step torque disturbance occurs such that there is an increase in the angle $\delta_{mf,r}$, with a subsequent increase in the field current reference $I^*_{f,r}$, and with $I^*_{m,r}$ initially fixed. The field current $I_{f,r}$ then begins a slow increase, with $I_{m,r}$ at first decreasing, and then to start increasing slowly. $I^*_{m,r}$ moves away from zero and a flux component thus is added to the stator current, which then changes the flux to attempt to compensate for the changing field current. At the end of the transient, the flux is at its initial value; however during the transient a fast change in the flux component is realized. Thus the machine returns to a unity power factor and $I^*_{M,r}$ returns to zero, with an improvement in response that is realized by using the field-oriented (i.e., vector) control. Similar performance is realized in the case of operation in the field weakening region, and thus the use of this technique is useful for the control of synchronous motors.

6.5 Motor and Drive Application

In general the synchronous motor is preferred, with a voltage source converter in the medium voltage range, and with a field-oriented controller. However, the cage rotor induction machine, or the cycloconverter, might be preferred in certain instances. Each of these has advantages and disadvantages, and thus a selection would require careful consideration based on the requirements of the actual application. The synchronous motor usually has the advantages of somewhat better efficiency than the cage rotor motor, but has higher initial cost and usually requires higher installation expense, plus others.

It is usually best, where applicable for the specific application that the motor and its drive are from the same manufacturer and provided as a single entity. Consultation with well-established, experienced, and reliable manufacturers of both motors and drives should be considered so that many of the more specific details can be thoroughly addressed. It is hoped that what was presented herein has helped to establish a useful basic initial insight to the application of motors and drives for tandem hot metal strip mills.

Appendix

Derivation of Angular Velocity of MMF Vector

In (6.3), (6.4), and (6.5) the peaks in the mmf in each winding of the stator are described, with ω being the sinusoidal frequency (rad/s) of the alternating current in the windings. It is assumed that the three-phase source is ideal, such that the peak current is the same in each of the phases, and that the number of turns is the same in each winding. Then the peaks of the mmfs are the same in each winding such that:

$$F_{\mathrm{a(max)}} = F_{\mathrm{b(max)}} = F_{\mathrm{c(max)}} = N I_{\mathrm{max}}, \tag{6.20}$$

where N is the number of turns in a winding, and I_{max} is the peak of the current in each phase. Then substituting into (6.2) gives:

$$\begin{aligned} F(\theta, t) = N I_{\mathrm{max}}\big[&\cos(\omega t)\cos(\theta) + \cos(\omega t - 120^\circ)\cos(\theta - 120^\circ) \\ &+ \cos(\omega t - 240^\circ)\cos(\theta - 240^\circ)\big], \end{aligned} \tag{6.21}$$

and using a trigonometric identity:

$$F(\theta, t) = \frac{3}{2} N I_{\mathrm{max}} \cos(\theta - \omega t), \tag{6.22}$$

which describes the rotation of the mmf wave around the air gap at the constant angular velocity ω. Thus at any time t_1, the positive peak of the electrical angle is at an electrical angle θ_1 ($= \omega t_1$), so that after $(t_2 - t_1)$ seconds the wave has traveled an electrical angle of $(\theta_2 - \theta_1)$. The rotating wave is also represented as a vector rotating around the air gap at an angular velocity ω.

Interaction of Flux Density Waves and Rotor MMF Waves to Produce Torque

As a beginning it is assumed that, in a cage rotor machine a balanced three-phase source is powering the stator, and that a separate machine is driving the rotor at the synchronous speed. In this case the following sequence can be noted:

- the rotating mmf wave produces a rotating flux density wave in the air gap;
- the rotating flux density wave is in phase with the mmf wave[25];
- since the rotor is being turned by the separate machine at the synchronous speed, there is no voltage induced in the rotor;
- the external machine is now quickly disconnected from the rotor;
- the rotor begins to reduce speed due to losses in the cage rotor machine;
- owing to the changing speed, the rate of change of flux in the rotor bars induces voltages in the rotor bars;
- the voltages in the rotor bars produce currents in the rotor bars;
- these currents produce an mmf wave that rotates at the same speed as the flux density wave;
- this interaction produces a torque;
- the torque balances the machine losses;
- since the losses are quite low, the rotational speed of the rotor is almost the synchronous speed;
- since the rotor is nearly at synchronous speed, the rate of change of the flux in the rotor bars is very low;
- because the rate of change of the flux is very low, the frequency of the induced voltages and currents is also very low;
- since the frequency is very low, the rotor reactance is very low, thus the rotor impedance is mostly resistive, so that the rotor current is nearly in phase with induced rotor voltages;
- thus the mmf produced by the rotor current lags the flux density wave by about 90° and is combined with the stator mmf wave;
- now since an external load torque is applied to the rotor shaft, the rotor slows to produce torque to balance the load torque;
- as the rotor slows, the frequency of the rotor current increases, the rotor reactance increases, and the angle β_r between the flux density wave and rotor mmf wave increases by δ;
- if the load torque further increases, the rotor reactance increases such that eventually the angle between the rotor mmf wave and the flux density wave will further increase; and at this point the motor will stall.

---

[25] Assumed that magnetic saturation effects are negligible.

References

1. Pittner J, Simaan MA (2011) Tandem cold metal rolling mill control, chap 6. Springer, London
2. Fitzgerald AE, Kingsley C Jr (1961) Electric machinery, the dynamics and status of electrotechnical energy conversion. McGraw-Hill, New York
3. McPherson G, Laramore RD (1961) An introduction to electrical machines and transformers. Wiley, New York
4. Sen PC (1967) Principles of electric machines and power electronics. Wiley, New York
5. Bose BK (2002) Modern power electronics and AC drives. Prentice-Hall PTR, Upper Saddle River
6. Pelly BR (1971) Thyristor phase-controlled converters and cycloconverters: operation, control, and performance. Wiley Interscience, New York

Index

J. Pittner and M. A. Simaan, *Tandem Hot Metal Rolling Mill Control*, Advances in Industrial Control, https://doi.org/10.1007/978-3-031-58082-6

GPSR Compliance

The European Union's (EU) General Product Safety Regulation (GPSR) is a set of rules that requires consumer products to be safe and our obligations to ensure this.

If you have any concerns about our products, you can contact us on ProductSafety@springernature.com

In case Publisher is established outside the EU, the EU authorized representative is:

Springer Nature Customer Service Center GmbH
Europaplatz 3
69115 Heidelberg, Germany

Batch number: 10732090

Printed by Printforce, the Netherlands